DATE DUE

BRODART, CO.   Cat. No. 23-221

Greenwood Guides to the Universe
Timothy F. Slater and Lauren V. Jones, Series Editors

Astronomy and Culture
Edith W. Hetherington and Norriss S. Hetherington

The Sun
David Alexander

Inner Planets
Jennifer A. Grier and Andrew S. Rivkin

Outer Planets
Glenn F. Chaple

Asteroids, Comets, and Dwarf Planets
Andrew S. Rivkin

Stars and Galaxies
Lauren V. Jones

Cosmology and the Evolution of the Universe
Martin Ratcliffe

THE SUN

David Alexander

Greenwood Guides to the Universe
Lauren V. Jones and Timothy F. Slater, Series Editors

GREENWOOD PRESS
An Imprint of ABC-CLIO, LLC

Santa Barbara, California • Denver, Colorado • Oxford, England

Library of Congress Cataloging-in-Publication Data
Alexander, David, 1963–
   The sun / David Alexander.
      p. cm. — (Greenwood guides to the universe)
   Includes bibliographical references and index.
   ISBN 978-0-313-34077-2 (hard copy : alk. paper) —
ISBN 978-1-57356-778-7 (ebook)
   1. Sun—Popular works. 2. Solar activity—Popular works.
3. Astrophysics—Popular works. 4. Astronomy—Popular works. I. Title.
   QB521.4.A44   2009
   523.7—dc22   2009006640

13   12   11   10   9   1   2   3   4   5

This book is also available on the World Wide Web as an eBook.
Visit www.abc-clio.com for details.

ABC-CLIO, LLC
130 Cremona Drive, P.O. Box 1911
Santa Barbara, California 93116–1911

This book is printed on acid-free paper ∞

Manufactured in the United States of America

To my parents, Cathy and David Alexander

Contents

Series Foreword   ix

Preface   xi

Acknowledgments   xiii

1   Introduction   1
2   A Shining Star   13
3   The Magnetic Sun   29
4   Looking Inside the Sun   43
5   Solar Acne: Sunspots and the Solar Surface   59
6   The Ever-changing Sun   79
7   The Many Faces of the Solar Atmosphere   99
8   Blowing in the Wind   117
9   Solar Storms   131
10 Space Weather   153
11 The Sun and Climate   173
12 The Future   195

Appendix: Interesting Facts about the Sun   207

Glossary   209

Annotated Bibliography   221

Index   223

Series Foreword

Not since the 1960s and the Apollo space program has the subject of astronomy so readily captured our interest and imagination. In just the past few decades, a constellation of space telescopes, including the Hubble Space Telescope, has peered deep into the farthest reaches of the universe and discovered supermassive black holes residing in the centers of galaxies. Giant telescopes on Earth's highest mountaintops have spied planet-like objects larger than Pluto lurking at the very edges of our solar system and have carefully measured the expansion rate of our universe. Meteorites with bacteria-like fossil structures have spurred repeated missions to Mars with the ultimate goal of sending humans to the red planet. Astronomers have recently discovered hundreds more planets beyond our solar system. Such discoveries give us a reason for capturing what we now understand about the cosmos in these volumes, even as we prepare to peer deeper into the universe's secrets.

As a discipline, astronomy covers a range of topics, stretching from the central core of our own planet outward past the Sun and nearby stars to the most distant galaxies of our universe. As such, this set of volumes systematically covers all the major structures and unifying themes of our evolving universe. Each volume consists of a narrative discussion highlighting the most important ideas about major celestial objects and how astronomers have come to understand their nature and evolution. In addition to describing astronomers' most current investigations, many volumes include perspectives on the historical and premodern understandings that have motivated us to pursue deeper knowledge.

The ideas presented in these assembled volumes have been meticulously researched and carefully written by experts to provide readers with the most scientifically accurate information that is currently available. There are some astronomical phenomena that we just do not understand very well, and the authors have tried to distinguish between which theories have wide consensus and which are still as yet unconfirmed. Because astronomy is a rapidly advancing science, it is almost certain that some of the concepts presented in these pages will become obsolete as advances in technology yield previously unknown information. Astronomers share and value a world-view in which our knowledge is subject to change as the scientific enterprise makes new and better observations of our universe. Our understanding of the cosmos evolves over time, just as the universe evolves, and what we learn tomorrow depends on the insightful efforts of dedicated scientists from yesterday and today. We hope that these volumes reflect the deep respect we have for the scholars who have worked, are working, and will work diligently in the public service to uncover the secrets of the universe.

Lauren V. Jones, Ph.D.
Timothy F. Slater, Ph.D.
University of Wyoming
Series Editors

Preface

A volume in the Greenwood Guides to the Universe Series, *The Sun* provides a basic introduction to one of the best-studied astrophysical objects in the universe. In some sense, the Sun is one of the most basic of cosmic entities, that is, a fairly average star, and yet our ability to observe it up close has shown it to be a highly complex system exhibiting a wide range of physical processes and phenomena. Understanding the physics governing the behavior of the Sun, therefore, has a powerful impact on our understanding of the physics at work in the universe. By observing the Sun, we now have a good idea of how magnetic field and plasma interact, how energetic particles produce the wide range of electromagnetic radiation observed, and how the dynamic motions of the interior are connected to the activity of the surface and the atmosphere. These are all processes that drive astrophysical phenomena throughout the universe.

The Sun is intended for a wide range of interested readers. The text is written with the nonspecialist in mind, but should also be useful to students and public library patrons interested in achieving a basic understanding of astronomy, astrophysics, and space physics. The volume comprises twelve chapters, each of which focuses on a particular facet of the Sun's behavior, the physical processes that govern this behavior, and the development of our understanding of these processes. The discussion takes the reader through the historical development of this understanding, from the early observational studies to the modern era where much of our information is obtained from state-of-the-art telescopes on space-based observatories. Sidebars in each chapter provide more in-depth technical information on a wide range of important and interesting topics. These sidebars are designed to present a deeper understanding without interrupting the flow of the main narrative. The physical concepts being discussed are augmented by a large number of observational images from telescopes around the world and in space.

Like all technical subjects, the study of the Sun has a language of its own, so to assist the reader we have provided a glossary of terms used in the chapters of this volume. Also, a single volume cannot do justice to the enormous wealth of multifaceted research on the Sun, nor to the sheer beauty and complexity evident in the literally millions of images and movies available from modern observations. In an attempt to make up for this lack, a number of additional print and electronic sources of information are provided at the end of each chapter and in a concluding general bibliography to allow the full impact of the modern study of the Sun to be appreciated.

Acknowledgments

My heartfelt thanks go to my family, Wendy, David, Andrew, and Caeleigh, for their patience and support during the writing of this book. I could not have done it without them. I would also like to thank my research group at Rice, particularly my students, for their perseverance in the face of my negligence when I had deadlines to meet. Holly Gilbert was of immense help via numerous discussions and a thorough reading of the draft manuscript. Her many questions and comments improved the content significantly. I would like to thank my colleague Phil Scherrer, who kindly gave me permission to use his quote about the Sun growing vegetables that opens the book and sets the tone for the Introduction and, in some ways, the whole book. I would like to acknowledge Michelle Larson as the originator of the radiative energy transport analogy given in Chapter 2 that she developed as part of the Yohkoh Public Outreach Project (YPOP) project. A good educational description never loses its use. I would like to thank my many friends and colleagues in the field of solar and heliospheric physics, without whom this book would not be possible. Their many discoveries, ideas, and visions make solar physics one of the most exciting and challenging areas of astrophysical research, and while no book is big enough to do this justice, I have tried to encapsulate in this volume, as best as I can, their enthusiasm and excitement. I would like to thank Lauren Jones for her comments on Chapters 1 through 4, and John Wagner of Greenwood Press for his review of the final manuscript. Finally, we would know next to nothing about the Sun if it was not for the high quality data provided by the ground-based observatories around the world and the space-based observatories circling the Earth and traveling the heliosphere. The hard work of the various instrument and observatory teams is very much appreciated.

1

Introduction

"The Sun is the only star known to grow vegetables." This pithy quote from Professor Phil Scherrer of Stanford University beautifully summarizes the fascination and the uniqueness of the star nestling on our doorstep. As stars go, the Sun is not something to write home about except for the fact that we would have no home without it. The energy provided by the Sun sets the conditions for the formation and development of life on Earth, life that eventually led to you, me, and the other 7 billion humans who inhabit this water-laden ball of dirt circling some 150 million kilometers from the star at the center of our solar system. Whether there is life on other balls of dirt circling other stars is, as yet, unknown. That there are planets orbiting other stars is now a matter of fact, with some 342 planets having been discovered to date lying in 289 distinct planetary systems. The planets discovered so far are more like Saturn and Jupiter in our own solar system—massive planets, probably gaseous, where life, if present, is most certainly very different from anything we have encountered on Earth. Until we can determine whether life exists or has existed on other planets in other star systems, we must assume that the Sun is a unique star that has the enviable characteristic of being able to grow vegetables.

While growing vegetables is clearly one way that influences how we humans view the Sun, throughout human history there has been a special reverence, both practical and mystical, for this daily visitor to our skies. Cultures around the globe have revered the Sun as a divine being and utilized the predictability of its annual wanderings through the stars to mark special times of year, relating its behavior to changes in the natural world around them. Celtic henges, Egyptian sun temples, aboriginal starlore,

Easter Island Moai, Native American medicine wheels, and many other ancient astronomical structures around the world mark out special points in the Sun's apparent path across the sky (e.g., solstices, equinoxes, and eclipses) and tell the tale of the universal importance that the Sun's patterns of motion had on the everyday life of our ancestors.

THE HELIOCENTRIC UNIVERSE

That the Sun is at the center of our solar system is an undisputed fact as we move into the twenty-first century. However, until the publication of *De Revolutionibus Orbium Coelestium* [Concerning the Revolutions of the Heavenly Spheres] by Polish astronomer Nicolaus Copernicus in 1543, a geocentric model of the solar system had dominated our ideas since the time of Aristotle (384–322 BC). The Earth-centered universe was supported by observation (at least, to the best of the abilities of the pre-telescope astronomers). Additionally, a model developed by Ptolemy (CA. AD 100–170) in which the planets wheeled around on numerous interlocking circular paths, known as *epicycles*, provided the backbone of the geocentric worldview. Such a complex mechanism of circles upon circles was a result of the dictates of Aristotle and his teacher Plato (CA. 428–348 BC), which stated that the Earth was at the center of the universe and that everything in the heavens was perfect and therefore unchanging, moving at constant speed in perfect circles. To match his detailed observations, Ptolemy had to resort to a complicated overlap of circular pathways.

While the Earth-centered solar system dominated, a number of other "theoretical ideas" were developed by the ancient Greeks, most notably Aristarchos of Samos (310–CA. 230 BC), who proposed the idea that the Sun lay at the center of the solar system and was seven times larger than the Earth (the true value is closer to 100 times). The dominance of Aristotle and Plato meant that these ideas lay essentially dormant for almost 1,800 years, until what is known as the *Copernican revolution*. It is not clear what drove Nicolaus Copernicus to consider a Sun-centered solar system, and, in fact, he was reluctant to publish his work—the first print of his book appeared on the day he died. It is often said that Copernicus did not like the complexity of the Ptolemy epicycle model and that this led him to the "simpler" picture of the planets revolving around the Sun. However, Copernicus was still constrained by the Aristotelian doctrines of circular and constant motion, and to make his Sun-centered model fit the observations, he also had to resort to epicycles. In the end, the Copernican model required more epicycles than Ptolemy's Earth-centered model (there is a debate over exactly how many each model required).

The Copernican ideas developed into a revolution of thought at the hands of German mathematician and astronomer Johannes Kepler (1571–1630), who devoutly believed in the Sun-centered universe proposed by

Copernicus. After decades of struggling to understand the motion of the planets around the Sun, Kepler eventually made a revolutionary breakthrough: he threw out the Aristotelian constraints of circular orbits and constant velocities. Kepler found that by allowing the planets to move in elliptical orbits, with the Sun at one focus of the ellipse, and the speed to vary with distance from the Sun, all the data could be explained. Kepler's three laws of planetary motion state the following:

- The orbit of each planet about the Sun is an ellipse with the Sun at one focus.
- As a planet moves around its orbit, it sweeps out equal areas in equal times. This is the equivalent to allowing the speed to change with distance from the Sun.
- More distant planets orbit the Sun at slower average speeds such that the square of the orbital period, p (in years), equals the cube of the average distance from the Sun, a (in astronomical units): $p^2 = a^3$.

These laws, based entirely on observations, eventually paved the way for Sir Isaac Newton's theory of gravitation.

THE SUN IN RELIGION AND MYTHOLOGY

From the earliest times, the Sun has featured in the mythology, religion, and cultural lives of people around the world. As the source of heat and light and the banisher of night, the Sun has had a preeminent role to play in the cultural and religious development of various societies throughout history. The Sun also marked out the passage of time, separating day from night, summer from winter, year from year. The combination of religion and timekeeping is intricately linked as many of the ancients believed that it was the power and mercy of the gods that led to bountiful harvests if proper sacrifice was made at the correct time of year. Failure to perform the required duties at the right time angered the gods and caused a failure of the crops or a devastating event, such as a plague of locusts or an early freeze. The progression of the seasons therefore had a strong religious aspect to it, and at the center of this was the Sun. Knowledge of the Sun's apparent motion across the sky and the times of highest and lowest points in the sky marked key dates in religious and agricultural calendars, and the solstices were of particular relevance. Celebration of the solstices persists to the modern day.

From the Greek god Apollo, whose golden chariot carried the Sun across the sky each day, to the Aztec Sun god Huitzilopochtli, who is considered an incarnation of the Sun, and the aboriginal Sun goddess Wuriupranili, who carried the flame across the sky from the east before dipping it in the western sea, evidence for Sun worship can be found around the world and across many civilizations and peoples through the ages. The henges (Figure1.1) of Stone-Age Europe, the medicine wheels of Native American

Figure 1.1 Stonehenge. Photograph taken on October 31, 2007, by Matthew Brennan.

culture, the Sun Dagger of the Anasazi People in Chaco Canyon, New Mexico, the Templo Major of the Aztecs, the city of Chitzen Itza of the Mayans, the Konarak Sun Temple in India, the Sun Temple of Rameses II at Abu Simbel in Egypt, and Macchu Pichu in Peru are among the many sites that focus on the Sun as a central god. All these religious sites share a common theme: the various alignments of the structures point to given astronomical events, of which the time of the solstice is one of the most prominent.

Chitzen Itza

One of the most impressive archaeological observatories is that of Chitzen Itza, the Mayan city in the Yucatan peninsula in Mexico. From about AD 550 to 800, Chitzen Itza was the ceremonial center of the Mayan civilization. The Maya were accomplished mathematicians and astronomers, as well as skilled farmers. Their agricultural success was, in part, a result of their aptitude at astronomy and vice versa. Two buildings at Chitzen Itza, El Castillo and El Caracol, are designed specifically with the Sun in mind. During the solar equinoxes, a snake, formed by the solar-cast shadows, descends the stairway of El Castillo. This is thought to be a manifestation of the Mayan god Kukulkán, the feathered serpent. The Maya could have used the patterns of shadows on El Castillo to track the seasons and to mark out significant times, like the equinoxes and solstices. At the observatory of El Caracol, narrow windows are aligned in such a way that at certain times of the year significant astronomical events can be viewed through the windows. A major focus of El Caracol is on the motion of Venus across the sky, but various alignments of both the windows and the buildings mark out the position of the solstices and equinoxes.

THE SUN AS A TIMEKEEPER

In addition to the importance of the yearly march of the Sun through the stars, the daily path of the Sun across the sky also led to important developments by the ancients, developments that we take very much for granted today. The regular pattern of behavior exhibited by the Sun creates a direct relationship between the Sun's predictable behavior and important times during the year (planting and harvesting being among the most important). This relationship added practical aspects to the normal religious observances and set the scene for the development of the calendar, the clock, and the science of astronomy. The Sun as a celestial timekeeper, and its impact upon the daily lives of humankind, makes it unique as an astrophysical object.

The Path of the Sun in the Sky

From our perspective on Earth, the Sun travels across the sky from east to west daily, with the shape and location of the path changing every day over the year. In reality, this motion is a combination of the Earth's eastward rotation, causing the Sun to rise in the East and set in the West, the Earth's annual revolution around the Sun, causing the Sun to appear against a changing background of stars, and the fact that the Earth's axis is tilted by 23.5° relative to the plane containing the Sun and the planets (the ecliptic plane). Particularly important is the tilt of the Earth's axis, as this is the reason for the seasons experienced on Earth.

As you watch the Sun throughout the year, you will notice that sunrise will occur gradually further north as the year progresses from winter to summer (in the northern hemisphere) with the Sun rising south of due east in the winter and north of due east in the summer. The time of maximum northerly deviation is known as the summer solstice and occurs typically on June 21 every year. Conversely, the furthest south point of sunrise occurs at the winter solstice (December 21). The Sun only rises due east on the equinoxes: spring (March 20/21) and autumnal (September 22/23). The 23.5° tilt of the Earth's axis creates four important latitudes with respect to the Sun's path: the Arctic and Antarctic circles and the Tropics of Cancer and Capricorn. The Arctic and Antarctic Circles lie at a latitude of 66.5° North and South, respectively, while the tropics lies at 23.5° North and South, respectively. The tropics are distinguished by the fact that the Sun is directly overhead on one of the solstices, while above the Arctic and Antarctic Circles the Sun is always above the horizon for six months of the year and below it for the other six months (see Figure 1.2).

The origins of our modern calendar date back to the ancient Egyptians, who were using a 365-day year as early as 4200 BC. The Egyptian calendar was based on the time between spring equinoxes (the tropical year), that is, the day at which the Sun crosses the equator on its way north. Because the tropical year is actually 365.25 days long, the calendar used by the ancient Egyptians would drift by one day every four years with respect to the seasonal changes. Over a period of one hundred years, the equinox would occur almost a month earlier. This phenomenon ushered in the concept of a leap year, where a day was added once every four years (February 29) to bring the calendar in line with seasonal variations.

- The Midnight Sun -
© Anda Berecsky, 2005

Figure 1.2 Time-lapse photograph showing the path of the Sun on midsummer's day from above the Arctic circle. Source: http://isilmetriel.deviantart.com.

The idea of keeping time daily also began with the ancient Egyptians about 700 years later around 3500 BC. The Sun's location in the sky was used to break the day into equal portions, and the changing day was marked out by a variety of means, progressing from simple shadow-casting by large obelisks to more elaborate and accurate shadow-clocks (or sundials). The shadows cast by the Sun also led the Greek mathematician Eratosthenes (296-194 B.C.) to determine, around 240 BC, that the Earth was round with a radius of 252,000 stadia, or 39,690 km (the true value is measured today at 40,008 km). Life in the ancient world moved more slowly than life today, but accurately telling time from the Sun paved the way for the development of navigational tools and improved astronomical observations. The use of the Sun as a navigational tool was of particular importance because it led to the "repeatable" exploration of the world by allowing travelers to navigate consistently between a given location and their home port—an obviously useful capability. Explorers used the height of the Sun in the sky (relative to the southern horizon, typically at noon) at any given time of year to determine their latitude north or south of the equator. Tools were developed to measure as accurately as possible the height of the Sun above the horizon, namely, the cross-staff (a version of this instrument was also used by astronomers), astrolabe, and sextant. Accurate determination of longitude, however, had to await the invention of an accurate chronometer.

While latitude could be determined relatively easily by measuring the height of the Sun above the horizon at a given time of day, it is significantly more difficult to measure one's longitude. The main reason lies in the fact that there is no natural reference frame, such as the horizon. The path of the Sun across the sky looks the same from any longitude; it is just delayed in time (e.g., San Francisco is three hours "behind" New York City) as the path is defined by the Earth's rotation. To measure one's longitude, one must know not only what time of day it is, but what time of day it is at a given reference point, for example, Greenwich in London. Local noon at a given point on the ocean occurs when the Sun is at its highest point in the sky. If one knew the time of day at the reference point, then the difference would tell you the longitude, since twenty-four hours marks out the complete 360-degree rotation of the Earth. To know the time at the reference point, one had to carry a clock that could keep accurate time over the several months of a voyage. It was the English clockmaker John Harrison (1693–1776) who won the British Navy prize of £20,000 (over US$12 million in today's money) by developing a chronometer that was accurate to about one-third of a second per day. Overall, the commerce and trade resulting from the development of navigational tools led to the wealth and power of a number of nations over the millennia (from the Phoenicians to the British). This further drove the development of better and more accurate astronomical measurements, and ultimately to more detailed observations of the Sun.

The Length of the Day

Our 24-hour day is also tied into the passage of the Sun. We normally define the length of a day by the time it takes a planet to rotate once about its axis. However, the Earth rotates in 23 hours 56 minutes and 4.1 seconds, slightly less than 24 hours. The 24-hour day is based on the time it takes the Sun to cross the sky, that is, the time between when the Sun is at its highest point in the sky on successive days. This is known as the *solar day*, and on average is 24-hours long with a variation of plus or minus 25 seconds. The reason the solar day is different from the rotation period of the Earth is because of the additional motion of the Earth around the Sun.

The tropical year is 365.242375 days, so in order for the spring equinox to occur at the same time every year, leap days are not added at a new century change unless the new century is divisible by 400, not just 4. This makes up for the roughly 1/100 difference between 365 and one-quarter days and 365.242375 days. In other words, the year 2400 will be a leap year, but 2500 will not.

THE SUN AND ASTROLOGY

The annual path of the Earth around the Sun means that the direction of the Sun in the sky is constantly changing. In fact, a person's birth sign (astrological sign) is determined from the constellation in which the Sun lay on the day he or she was born. The distribution of stars in the nighttime sky led to early civilizations developing a system of grouping the stars into patterns known as constellations, many of them representing some mythological being. A special group of constellations are those that lie along the path of the Sun and are known as the zodiac; thus the Sun appears to be in the zodiacal constellation of Leo on August 21. Traditionally, there are twelve zodiacal constellations (one for each month of the year), and depending upon when you were born, the position of the Sun would decide your birth sign. This laid the basis for the practice of astrology where people's personalities are determined by their birth signs and their futures are governed by the position of the planets and the Sun in the sky. Unlike astronomy, astrology is not based on any physical principles and does not allow for the ever-changing skies.

If you look carefully at a star chart, you will see that the Sun actually passes through thirteen constellations, with Ophiuchus, the Serpent Bearer, lying between Scorpius and Sagittarius. While Ophiuchus is not considered a part of the western astrological zodiac, which is based on the division of the Sun's path into twelve equal periods, it is included in the zodiac used by Hindu astrologers. At present, the Sun is in Ophiuchus from November 30 to December 17.

Another physical behavior not accounted for in astrology is the rotation of the Earth's axis, which over the 1,700 or so years since the establishment of the zodiacal constellations has resulted in a shift of the sky relative to the solar system. Currently, the Earth's axis points to Polaris (the North Star) but the axis has a wobble as it spins, just like a spinning top, and this wobble causes the axis to rotate to a different point in the sky, wobbling back to Polaris every 26,000 years. This means that the positions of the constellations relative to the Earth are constantly changing, and over the course of 1,700 years the Sun's path has "shifted" around the zodiac. In the present age, the Sun appears in the constellation Pisces on March 21, but anyone born on March 21 who wanted to check their horoscope would look under Aries.

THE SUN AS A STAR

In the modern world, understanding the patterns exhibited by the Sun is no less important, although the focus is somewhat different. Since Galileo (1564–1642) first identified sunspots as being solar in nature, our perception of the Sun as a perfect unchanging heavenly body was permanently altered. The march of the sunspots across the disk of the Sun, their changing numbers and sizes, and their patterns of appearance demonstrated that the Sun is variable on many different scales and led the way to modern solar physics. In the twenty-first century, our reliance on advanced technology for global telecommunications, navigation, and weather monitoring makes us susceptible to the vagaries of the Sun's behavior. The pattern of growth and decay in the number of sunspots over an eleven-year period (the solar cycle) and its related production of energetic activity both at the Sun and at the Earth defines our modern approach to observing the Sun.

Like the study of all astrophysical objects, observing the Sun relies on the detailed knowledge of how physical systems interact to produce light and other electromagnetic radiation. The Sun is approximately 150 million kilometers away, which is extremely close, compared to any other star, but not close enough to directly measure the physical makeup of its constituent parts. We do not have the luxury of being able to stick a thermometer into the solar atmosphere to measure its temperature, or to shine light through it to see how much is absorbed or transmitted. In other words, we can only measure what reaches us at the Earth, or, in some special cases, what is detected in interplanetary space by the various scientific spacecraft we have launched. All our information about the Sun, therefore, comes from the radiation it produces, whether it is X-ray photons from the million-degree corona, energetic particles accelerated by solar storms, or optical light from the Sun's surface. (Solar scientists refer to the sharp edge of the Sun that we see in telescopes as the surface of the Sun. This will be explained in more detail in Chapter 5.)

Distance Measures in the Universe

The distances between stars in the Milky Way, and the galaxies from one another, are so vast that if we were to use our usual units (kilometers or miles) we would spend a lot of time writing very large numbers with lots of zeros. To prevent this problem, astronomers adopt more convenient units of measurements. Imagine if you had to tell your friends how tall the Empire State building is in millimeters instead of feet, or how big the Pacific Ocean is in inches instead of miles. Pretty soon you would find yourself saying, "Look lets just say that there are 63,360 inches in a mile," and start using miles instead of inches. Astronomers do something similar when discussing the distance between stars, or galaxies. We define a unit based on how far light travels in one year, and with a distinct lack of imagination, call this a light-year; despite its name, a light-year is a distance (*not* a time). Light travels at 299,792,458 meters per second, so in one year a beam of light would travel 9.4607×10^{12} km (or approximately 9.5 trillion kilometers). This distance is equivalent to one light-year. The nearest star to the Sun, Proxima Centauri, is 4.2 light-years away, or almost 40 trillion kms. If this sounds big, the size of the observable universe is some 28 billion light years across.

Today, we observe the Sun from advanced observatories around the world and from space-borne telescopes orbiting the Earth and traveling through the solar system. The telescopes and instruments in these observatories and spacecraft observe the Sun at all wavelengths from the lowest energy, longest wavelength, radio waves to the highest energy, shortest wavelength, gamma rays. From the instruments in space, we also directly detect particles (protons, electrons, and ions) emanating from the Sun, allowing us to directly probe the makeup of the solar wind (see Chapter 8) and the nature of the particles accelerated in solar flares and coronal mass ejections (see Chapter 9). Novel techniques allow us to use these observations to probe the interior structure of the Sun (in much the same way that we use earthquakes on the Earth to learn about the composition of the terrestrial interior) to determine the properties of the magnetic field of sunspots, to infer the temperature and density of the different parts of the solar atmosphere, and to measure the speeds and directions of the motions of the solar plasma. The detailed analysis of all of these observations has led to a fairly complete picture of how the Sun works, how its different parts interact, and how it affects the Earth. The description of this picture is the main focus of this volume.

Despite all the information we have garnered since Galileo about the physical workings of the Sun, a number of mysteries remain. We don't know why the Sun's corona, at one million Kelvin (K), is so much hotter than its photosphere, at a mere 5800 Kelvin. (Kelvin is used in physics and astronomy to measure temperature—water freezes at 273 K, boils at 373 K, and 1 K is the same as one degree centigrade but its scale starts at absolute zero.) We don't know what causes the violent release of huge amounts of energy from the magnetic field in a solar storm, heating the solar atmosphere to tens of millions of degrees, accelerating particles to relativistic energies, sending large volumes of plasma hurtling out into space at several thousand kilometers per second, and producing enhanced radiation across the electromagnetic spectrum. We don't know how the Sun's magnetic field is generated in the solar interior to make its way to the surface and beyond where it dominates solar activity. We don't know why this magnetic field varies over an eleven-year period, generating a regular change in the number of sunspots, and consequently the number of solar storms. We don't know why this regular cycle is sometimes interrupted to produce long periods of reduced sunspot numbers with noticeable climatic effects on the Earth.

The fascination of the Sun as an astrophysical object lies in the answers to these "don't knows." Understanding how the Sun works, from the energy production in its core to the generation of magnetic field and the impact on its surroundings, has important consequences for our understanding of astrophysics in general, and for our understanding of how other stars may interact with their planetary systems and provide the conditions for growing vegetables on other worlds. Our ability to probe, explore, and examine the Sun in detail now makes the Sun an ideal laboratory for astrophysics. As we gain in knowledge and understanding of this star next door, we gain critical insight into how stars work, how plasma and magnetic field interact in astrophysical systems, and how stellar variability influences the development of life on planets. This volume provides an in-depth look at the Sun, from the generation of energy in the solar core to the interaction of the solar atmosphere with the Earth. Throughout the volume we discuss how modern instrumentation enables us to build a complete picture of the Sun, how the different parts of the Sun interact, what impact solar variability has on the solar system, and what progress we can look forward to in the future.

RECOMMENDED READING

Gould, Alan, Carolyn Willard, and Stephen Pompea. *The Real Reason for Seasons: Sun-Earth Connections*. Great Explorations in Math and Science (GEM). Berkeley: Lawrence Hall of Science, University of California, 2000. (http://lawrencehallofscience.org/gems/GEMSSeasons.html).

Koeslter, Arthur. *The Sleepwalkers: A History of Man's Changing Vision of the Universe*. London: Penguin, 1990.

Sobel, Dava. *Longitude: The True Story of a Lone Genius Who Solved the Greatest Scientific Problem of His Time*. New York: Walker & Company, 2007.

WEB SITES

Ancient Solar Observatories: http://www.traditionsofthesun.org.

Chitzen Itza: http://www.exploratorium.edu/ancientobs/chichen/flash.html.

JPL Exoplanet Web site: http://planetquest.jpl.nasa.gov.

Solar Week: http://www.solarweek.org.

Sun in World Culture: http://solar-center.stanford.edu/folklore.

2

A Shining Star

The Sun is a star—a simple but frequently forgotten fact. When we look into the night sky at the literally billions of stars that make up our galaxy and the billions of galaxies that make up our universe, we feel a sense of awe at the sheer vastness of space and the sheer number of potential worlds surrounding those stars. We create myths and fantasies about traveling to other worlds and what we might see there. We may even pay $39 to have one of them named after us or a loved one. Yet, when daylight comes and the sky is dominated by a single star, so bright that the other stars fade into invisibility, the wonder dissipates and is replaced by the daily humdrum that is the routine march of the Sun across the sky. The importance of the Sun in our lives, celebrated through the ages by cultures around the world (see Chapter 1), has been all but forgotten as modern advancement has brought us electronic replacements for keeping time and finding our way, the ability to cultivate our plants year round, and the demotion of the Sun from a god to a source of skin cancer. As we have learned more about the Sun as a star and characterized its behavior in the mathematical language of physics, we have lost the sense of its importance to us and how much we rely on it for our existence.

Yet, the importance of the Sun to our modern world cannot be overemphasized. We still owe our very existence to the heat and light it emits and the stabilizing gravity it provides, but our modern-day reliance on all things electronic has given us a much more direct connection to the Sun than we might at first imagine. The Sun and the Earth are in constant contact through the interaction of a fast-flowing plasma wind of ions and electrons (see the solar wind, Chapter 8). The impact of this wind on the Earth serves to generate an array of electromagnetic phenomena such as aurorae (the northern and southern lights), geomagnetic storms, and ionospheric disturbances. These electromagnetic phenomena in turn disturb the electrical environment of the Earth, causing electrical blackouts, enhanced radiation at high altitudes, disruptions and sometimes downright failures of the Earth orbiting satellites that provide much of the modern world's telecommunications (cell phones, satellite radio and television), weather and environmental monitoring, and security surveillance. This Sun-Earth interaction is collectively known as space weather and is covered in detail in Chapter 10. Understanding the Sun as a star, its variability, and its ability to generate magnetic fields is crucial to protecting our modern space resources, ensuring the safety of astronauts and high-altitude flight crews, and understanding the impact of this variability for life on Earth.

Beyond the practical aspects of everyday life in the twenty-first century, an understanding of the Sun as a star and the physics that governs its variability on short and long timescales provides insight into similar processes around the universe. The Sun has often been described as a laboratory for astrophysics, a laboratory where the experiments are uncontrollable but where their effects can be observed in unprecedented detail. Observing the physics of an astrophysical body close at hand allows us to infer what might be going on in similar environments throughout the universe, what physical principles are responsible for what observed characteristics, how plasma and magnetic field interact to produce heat and light, and, generally, understand one of the most important constituents of the universe: the star. This chapter explores what we know about the Sun as a star and what it tells us about our own place in the universe.

THE SUN AS A STAR

As has often been quoted, the Sun is a relatively mediocre, middle-aged, star with very few distinguishing characteristics, lying in the unfashionable neighborhood of the outer spiral arm of the spiral galaxy known as the Milky Way. The Sun lies approximately 28,000 light-years from the galactic center and takes a leisurely 250 million years to perform a complete orbit of the galaxy.

The consideration of the Sun as a star is a relatively recent idea dating back, as far as we can tell, to the famous mathematician and philosopher René Descartes (1596–1650). In his book *Principia philosophiae* published in 1644, Descartes suggested that the Sun was one of the panoply of stars that filled the night sky, all of which formed at the center of a vortex at the beginning of the universe. The idea of the vortex followed on Galileo's relatively recent identification of sunspots as being solar phenomena rather than an effect in the atmosphere of the Earth and that the observed motions of the sunspots indicated that the Sun rotated.

The study of the Sun as a star, and indeed the study of stars in general, took off with the invention of spectroscopy as an observational tool. The word *spectroscopy* comes from the word *spectrum*, which was introduced by Sir Isaac Newton in 1666 to explain the continuous spread of colors caused by the dispersal of sunlight by a prism. Spectroscopy of stars, however, did not really take off until the work of German physicist Joseph Fraunhofer (1787-1826) in the early 1800s. By spreading the light from the Sun over a large distance, he saw that the normal rainbow of colors was broken up by a large number of dark lines (named the Fraunhofer lines in honor of their discoverer). These were the first spectral lines ever observed and allowed great advances to be made both in the science of spectroscopy and our knowledge of stars, essentially founding the science of astrophysics (Figure 2.1). Fraunhofer made many refinements to the instruments he used to observe the solar spectrum, ultimately enabling him to measure the exact wavelengths of the dark lines. He did not, however, understand what caused them. The key was in an observation made in 1848 by French physicist Jean Foucault who observed that a flame containing the element sodium would absorb the yellow light emitted by a bright arc formed between two carbon electrodes. Previous work using spectroscopy in the laboratory had shown that different sources emitted bright lines of varying colors (*emission lines*), but the observation of Foucault demonstrated for the first time that dark lines at varying wavelengths (*absorption lines*) could be generated.

All of this information was finally brought together in 1859 by German physicists Gustav Kirchhoff (1824-1887) and Robert Bunsen (1811-1899). These scientists performed a number of experiments that demonstrated that each gas had its own unique spectrum (see sidebar on stellar fingerprinting). Kirchhoff also made the discovery that, at a given wavelength, the power emitted and the power absorbed are the same for all objects at the

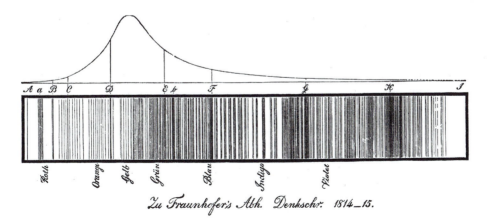

Figure 2.1 Joseph von Fraunhofer's (1787–1826) diagram of the lines of the solar spectrum. (Photos.com)

Scientific Notation

The numbers used in astronomy to describe distances or sizes of objects are so large that astronomers use a specialized notation to denote these large scales. The notation adopted is called *scientific notation* and uses an exponent to describe the powers of ten that need to be applied to the number given. For example, the number 10^{24} means a "1" followed by twenty-four "0's," while 4.7×10^9 is 4.7 times one billion (a "1" followed by nine "0's"). An interesting property of this notation is that to multiply two numbers, simply multiply the digits and add the exponents: $(3.5 \times 10^{17}) \times (4.2 \times 10^{13}) = (3.5 \times 4.2) \times 10^{30} = 14.7 \times 10^{30} = 1.47 \times 10^{31}$.

same temperature (this is known as *Kirchhoff's Law*). Thus, a gas that radiates a given emission line must, at the same temperature, absorb radiation at the same wavelength as the emission line. The Fraunhofer lines in the Sun's spectrum could now be explained as the absorption of specific wavelengths by the different elements making up the solar surface. A careful study of these absorption lines made the study of the solar atmosphere, and that of other stars, possible. The techniques of Fraunhofer, Kirchhoff, and Bunsen are still used today to analyze the chemical composition of astrophysical objects. Stars are now distinguished by their spectral type. The Sun is officially classified as a yellow main sequence star of spectral type G2. It is interesting to note that in 1835 the famous French philosopher August Comte emphatically stated that no matter what advancements in science man achieved, he would never be able to know the composition of stars.

Stellar Fingerprinting

A quick scan of the night sky will show that stars are different colors; for example, look at the shoulder and knee of Orion to see the red supergiant star Betelgeuse and the blue-white supergiant star Rigel, respectively. The color of a star is a measure of the surface temperature of the star, and through careful spectroscopic studies we can also learn about the chemical elements that are present. The German physicist Gustav Kirchhoff first established that each chemical element, compound, and molecule has a distinctive pattern of wavelengths that it can absorb or emit, that is, the element has a specific fingerprint. The combination of lines in a star's spectrum, its spectral type, is used to compare stars with one another and to classify them in groups. For example, the bulk of the Sun's light is yellow, so it is classified as a yellow dwarf star, or in the terminology of astronomers, a G star. A further refinement based on its temperature makes it a G2 star. For comparison, Betelgeuse is classified as an M2 star and Rigel as a B8 star. The spectral sequence from hot (>30,000 K) to cool (<3,000 K) has the letter designations OBAFGKM, so O and B stars are hot and G, K, and M stars are cool. (For more detail on the spectral classification of stars, see the volume in this series called *Stars*.)

An approximate age for the Sun is determined from the age of some of the oldest objects in the solar system—meteorites. Prevailing theories assume that the solar system collapsed from a single gaseous nebula, with the Sun forming slightly ahead of the planets, asteroids, and comets. Objects such as asteroids and comets, parts of which rain down on us as meteorites, are effectively unchanged since the solar system formed and therefore can be used to determine how long the solar system has been around. The age of meteorites found on Earth is determined from the decay of radioactive isotopes such as Rubidium-87, which decays into the stable, and therefore fixed, isotope Strontium-87. (To learn more about meteorites in the solar system, see the volume in this series called *Asteroids, Comets and Dwarf Planets*.)

The typical timescale for half the number of Rubidium-87 atoms to decay into it's daughter product, Strontium-87, is 4.8×10^{10} years. By measuring how much Rubidium and Strontium are present in a meteorite, we can determine how long ago the meteorite was formed, and therefore a

Measuring the Mass of the Sun

Most schoolchildren know that the Sun is 93 million miles (150 million kilometers) distant from the Earth. However, not many realize that we have only known this number for less than 150 years. Newton's laws of gravity, first appearing in the publication of his treatise *Philosophiae naturalis principia mathematica*, commonly known as *Principia*, in 1687, led to great advances in the understanding of planetary motion. However, with no accurate measurement of the true distance from the Earth to the Sun, astronomers could only determine the distances and orbits of other planets in the solar system in terms of the Earth's distance from the Sun (1 Astronomical Unit or 1 AU) and the Earth's one-year orbit. It was almost seventy-five years later that careful observations by many dedicated astronomers who had ventured to the remotest parts of the globe led to a determination of how big 1 AU actually was in physical units (miles or kilometers). By observing the transit of Venus across the Sun simultaneously from several different latitudes, they determined the value of 1 AU within the range of 1 AU = (149–156) million kilometers. Venus transits happen in pairs separated by 8 years with each pair being separated by 105–122 years (the most recent transit occurred on June 8, 2004, the next one will be June 5, 2012). The rarity of transits meant that it was very important to use the ones that came around as effectively as possible. The Venus transits of 1761 and 1769 gave the first opportunity to perform the necessary observations with sensitive enough telescopes to attain a value for the astronomical unit. For the astronomers of the eighteenth century, it would be the observation of a lifetime. Subsequent transits refined the value of the astronomical unit, but it wasn't until the Venus transit of 1882 that the modern value of 93 million miles was attained with sufficient precision, some 200 years after the original publication of Newton's *Principia*.

The importance of this major breakthrough was that, once the physical scale was determined, the relationship between orbital period and distance in Newton's Law of Gravitation allows one to calculate, from observable quantities, the mass of the Sun. The square of the period of a planet's orbit is proportional to the cube of the distance to the Sun divided by the mass of the Sun (this assumes that the planet's mass is negligibly small compared to that of the Sun). Knowing the period in seconds and the distance in kilometers means that this relationship can be solved for the mass of the Sun, which turns out to be approximately 2×10^{30} kg.

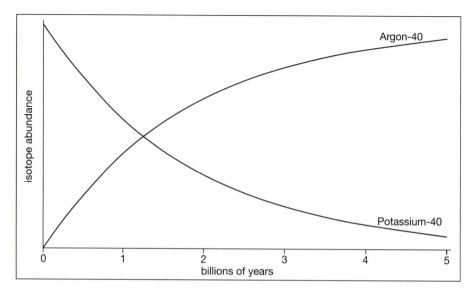

Figure 2.2 Radioactive decay of potassium-40 into argon-40. The exponential decay shows a half-life of 1.25 billion years. Illustration by Jeff Dixon.

lower estimate of the age of the Sun (Figure 2.2). These measurements give the age of the Sun to be at least 4.52 billion years old (plus or minus 30 million years). Sophisticated computer models of sun-like stars give a higher age of 4.57 billion years, consistent with the meteorites being younger than the Sun. Table 2.1 summarizes the vital statistics that serve to describe the Sun as a star, some of which will be of more use than others in the forthcoming chapters.

Other important numbers relating to the Sun are:

- Average Earth-Sun Distance: 149.6 million km (92.6 million miles)
- Elemental Composition of Sun: 74% hydrogen, 25% helium, 1% other (by mass).
- Luminosity: $\sim 4 \times 10^{26}$ Joules/s. [As bright as 4 trillion trillion 100-Watt light bulbs]

Table 2.1. **The defining characteristics of the Sun as a star**

	SUN	EARTH	RATIO OF SUN TO EARTH
Radius (km)	696,000	6,376	109
Mass (kg)	1.988×10^{30}	6×10^{24}	332,946
Volume (m³)	1.41×10^{27}	1.1×10^{21}	1.3 million
Average density (kg/m³)	1408	5506	1/4
Surface gravity (m/s²)	273.95	9.81	27.9
Rotation period (days)	26 (at equator)	1	1/26
Temperature at surface (K/°C)	5785/5512	~293/~20	275.5
Escape velocity at surface (km/hr)	2.223 million	40,228	55.26

THE PRODUCTION OF SUNLIGHT

Clearly, what makes the Sun a star is that it shines. The Sun produces its own energy, in the form of heat and light, which it "shares" with the rest of the solar system. A fundamental physical law of nature is that energy cannot be created nor destroyed, known as *conservation of energy*. However, energy can change form from one type to another. Rub your hands together and they get warm—you are converting biochemical energy in your arm muscles, which they get from the food and drink you ingest, to kinetic energy in the motion of your hands, which in turn transfer this energy to thermal energy via friction between your two hands. This means that the heat and light, radiant energy, produced by the Sun must be the result of some energy conversion process. Discovering how the Sun, and other stars, produce the vast amount of energy required to keep it burning for at least the 4.5-billion-year lifetime of the Earth was one of the fundamental land-marks of twentieth-century astrophysics.

As recently as one hundred years ago, the source of the Sun's energy was still a great topic of debate among scientists. Very little progress had been made on this topic since the Greek philosopher Anaxagoras (CA. 500–428 BC) suggested that the Sun was a hot, glowing rock no bigger than an aver-age Greek island. By the end of the nineteenth century, the distance to the Sun had been calculated to pretty good precision (see sidebar) enabling similar precision in estimating the physical size of the Sun. Using measure-ments of the energy output of the Sun at the Earth, astronomers were then able to calculate the total amount of radiant energy emitted by the Sun.

An experiment first performed by German-born British astronomer Sir William Herschel (1738–1822) illustrates how much radiation is being emitted by the Sun. He measured how long it would take to completely melt a thin layer of ice sitting in direct sunlight. The time taken, the volume of the ice melted, and the specific thermal properties of the ice allows one to estimate how much radiant power reaches the surface of the Earth from the Sun (see calculation). Even at the great distance of the Earth from the Sun, the ice will melt in a matter of minutes (approximately fifty-five minutes for one cubic centimeter of ice). From this information, Herschel calculated that the Sun bathed the surface of the Earth with approximately 1,000 Watts of power in every square meter. Allowing for the effects of the atmosphere, the Sun's actual power at the distance of the Earth is 1,366 Watts/m^2—this is known as the *solar constant*. The Earth only intercepts a small fraction of the total radiation from the Sun. Translating the numbers from Herschel's experiment to the whole Sun, we find that the Sun radiates so strongly that if the entire amount of energy radiated was counted, it could melt a volume of ice the size of the Earth at a distance of 150 million kilometers in about 15.5 minutes.

The Sun produces enough energy in one second to power all of the United States for over a million years! This vast amount of energy had to be

....

The Solar Constant

To calculate how much sunlight arrives at the Earth's surface, we can measure how long it takes to melt a circular piece of ice 1 m in diameter and 0.5 cm thick. To do this we need to know a few numbers.

- Earth-Sun distance: 1.5×10^8 km
- Specific latent heat of ice: 3.34×10^5 J/kg (i.e., how much energy to melt 1 kg of ice)
- Density of ice: 1000 kg per cubic meter
- Time to melt ice: 28 mins = 1680 s

Volume of ice = area × thickness
$$= (\pi \times 0.5^2)\text{m}^2 \times (0.5 \times 10^{-2})\text{m}$$
$$= 3.927 \times 10^{-3} \text{ m}^3$$

Mass of ice = volume × density = 3.927 kg

Energy to melt ice = mass × latent heat = 1.312×10^6 J

Area of ice = $(\pi \times 0.5^2)\text{m}^2 = 0.785$ m^2

Power per unit area at surface of Earth from Sun = energy ÷ area ÷ time
$$= 1.312 \times 10^6 \div 0.785 \div 1680$$
$$= 994.55 \text{ W/m}^2$$

This value changes with time of day and year, as well as weather conditions. In space, the value is relatively constant and has been measured to be 1366 W/m^2 at the Earth's distance from the Sun.

....

accounted for by some plausible energy conversion process. Many suggestions and calculations were made by astronomers through the ages (see sidebar), but it wasn't until the discovery of nuclear processes and Einstein's famous equation, $E = mc^2$, that the true nature was discovered. The only source of energy sufficient enough to produce these large power values is the energy released in the fusion of atoms at the core of the Sun.

....

Powering the Sun

In the late 1800s, the main source of energy powering the industrial revolution was the burning of coal to produce electricity, steam, and so forth. So, scientists calculated how long the Sun would shine, at the measured rate, if the heat and light were produced by the burning of a Sun-sized, high purity, coal fire, that is, the conversion of chemical energy in the coal to radiant energy. Typically, coal produces approximately 24 million Joules of energy for every kilogram. If the Sun was entirely made of coal and was shining at its current luminosity, 4×10^{26} Joules per second, it would shine for a total of about 7,500 years. This went against the known geological age of rocks on the Earth, which were billions of years old, and so the coal-fired Sun was dismissed as implausible.

Another possible source of energy conversion was to hypothesize that the Sun was decreasing in size. This creates energy by converting gravitational potential energy into heat and light: think of how a dropped ball gains speed. A shrinking Sun would keep the Sun shining by continually generating heat and light as it got smaller. The Sun is not observed to be shrinking, so you would think this would not work. However, the mass of the Sun is so large that it would only need to contract a very tiny bit each year to create the necessary luminosity. The amount that the sun would shrink is so tiny that it would not be observable. Calculations show that if the Sun was emitting energy at its current levels via the conversion of gravitational potential energy, then the Sun would shine for up to 20 million years. This number is much larger than that derived from burning coal, but still woefully short of the 4.5 billion years required for the age of rocks on Earth.

It wasn't until the emergence of Albert Einstein and his famous equation, $E = mc^2$, that the true nature of the energy production in the Sun would become apparent. We now know that the Sun is powered by the conversion of mass to energy in the core, where under extreme densities and temperatures hydrogen nuclei are converted to helium nuclei through the process of nuclear fusion.

Modern theories state that the Sun and the solar system formed from the gravitational collapse of a large cloud of gas known as the *solar nebula*. As this cloud collapsed, the central region that was later to form the Sun got denser and denser and began to heat up, ultimately forming a large ball of hot gas with a very dense, very hot core. The intense heat in this core was so strong that the atoms there, mostly hydrogen, could not maintain their structure and were broken up into free protons and electrons that moved around extremely quickly (they had a lot of thermal energy). The core of the Sun is also a region of extremely high density, so these fast-moving particles cannot go very far before they hit other particles moving just as fast. It is the fusion, or slamming together, of these particles that creates the nuclear reactions that provide the energy that causes the Sun to shine.

Getting two protons to stick together is not that easy since they both have a positive charge and like charges repel. However, in the core the protons are moving so fast and collide with such violence that they can overcome the repulsion force of the similar charges to get close enough to stick together via a force that particle physicists call the strong force. Try pushing the similar poles of two magnets together. Push them gently and they deflect. Push them hard and you can make them touch. You exerted enough force to overcome the natural repulsion. When two protons stick together they produce an isotope of hydrogen called deuterium, consisting of a proton and a neutron; the physics of the interaction results in the conversion of a proton to a neutron in the new atom's nucleus. These deuterium atoms do not last long, but the collisions are so frequent that before they can break up, a large number of them collide with another free proton to form a new element, an isotope of helium with two protons and one neutron in its nucleus. This isotope is called helium-3. Two of these helium-3 ions then collide with each other to form "normal" helium (helium-4), which consists of two protons and two neutrons. The last interaction also releases two protons into the general mix, converting a total four protons into one helium-4 ion. The interesting thing is that one helium-4 ion has slightly less mass than the four protons that went into making it. It is this slight reduction in mass that provides the key to the production of energy in the Sun's core. A helium-4 ion is 0.7% less massive than four protons.

One of the most famous equations of all time is the one developed by Albert Einstein in 1905: $E = mc^2$, which states the relationship between mass and energy. In this equation, E is energy, m is mass, and $c = 3 \times 10^8$ m/s is the speed of light in a vacuum. In words, the energy associated with the mass of an object is equal to the mass times the speed of light squared, and since the speed of light is such a large number, a small amount of mass can, in principle, be converted into a lot of energy. Normally, mass does not spontaneously convert to energy. However, in the extreme conditions of the core of a star, the violent interactions between atomic nuclei cause mass and energy to essentially be the same thing. Thus, the 0.7% of mass that goes astray in the production of helium from hydrogen is not lost, but converted to energy. In all, some 600 billion kg of hydrogen is fused into helium *every second* in the center of the Sun. The mass converted to energy is then simply 0.7% of this mass, which is approximately 4.2 billion kg. Using Einstein's equation, this means that every second the Sun's core produces approximately 3.78×10^{26} Joules of energy. Thus, in one second the Sun produces enough energy to power the United States for one million years.

THE STRUCTURE OF THE SUN

The structure of the Sun is essentially defined by two physical processes: the production of energy by nuclear fusion in the core and the transfer of that energy to the surface. The density and temperature of the different parts of the Sun are then determined by the outflow of energy from the core and the effect this energy flow has on the atoms and ions that make up the solar interior and atmosphere. Although the Sun is completely made of gas, the density and temperature of the gas changes drastically from the center to the outermost regions. Models of stellar evolution for Sun-like stars suggest that in the core of the Sun the density can be as high as 1.5×10^5 kilograms per cubic meter. In other words, a piece of the Sun's core the size of a tennis ball would have about as much mass as a National Football League linebacker. At the other extreme, near the base of the outermost layer, the corona, the density drops to about 1.0×10^{-12} kilograms per cubic meter. This value is close to laboratory vacuum densities on Earth. The models also suggest that the temperature of the core is about 14 million Kelvin and slowly cools out to the solar surface. The luminosity of a star is extremely sensitive to the temperature of the core, and so the core density and temperature implied from the models are believed to be very accurate. At the solar surface, the spectroscopic studies indicate that the temperature is approximately 5,800 Kelvin. As we move into the solar atmosphere, the temperature starts to rise again, reaching 1–2 million Kelvin in the solar corona. In the corona, emission lines of highly ionized iron ions indicate that the temperature is extremely high (see Chapter 7 for a full discussion). This is puzzling since one would expect that the further we are from the source of energy, in this case the Sun's core, the colder it should get. This holds true in the solar interior, but the rise in temperature above the solar surface presents a major puzzle to solar scientists. We know that it is related to the presence of magnetic fields on the Sun, but the details remain elusive. We will revisit this problem in Chapter 7.

The energy created by the nuclear fusion in the core then has to work its way out to the surface for the star to officially "switch on." The physical transport of energy to the stellar surface for a star like the Sun is done in two distinct ways (Figure 2.3). The first energy transport process, which takes the energy produced in the core through the solar interior up to about 70% of the way to the surface, is via the absorption and emission of radiation; this internal region surrounding the core is known as the *radiation zone*. Atoms near the bottom of the radiation zone absorb the radiation

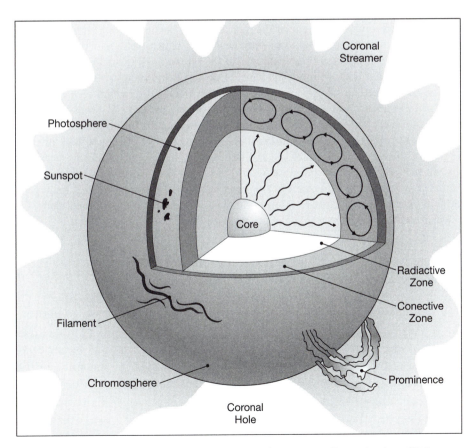

Figure 2.3 Cutaway of the Sun showing the internal structure. Illustration by Jeff Dixon.

resulting from the fusion process and become "excited," storing the absorbed energy for a short time before "relaxing" back to their previous state by emitting the excess energy as new radiation. This new radiation is then passed among the atoms throughout the radiation zone, transferring heat upwards through the Sun. The stellar evolution models indicate that the temperature in the radiation zone is a little cooler than at the core, with the temperature falling from 15 million Kelvin at the base to around 2 million Kelvin at the top.

Solar Neutrino Problem

A side product of the nuclear reactions in the Sun's core is the production of small elementary particles called *neutrinos*. Neutrinos do not interact very well with regular matter, so all of the neutrinos produced zoom quickly through the Sun and spread outwards into space. Right now there are about a thousand trillion neutrinos passing through the human body every second or so. If we could detect these neutrinos, and count their number, we could "see" directly into the Sun's core and study the details of the nuclear fusion process. The fact that the neutrinos pass through almost everything makes them very difficult to detect. However, neutrinos do occasionally interact with matter, via collisions with the atomic nuclei, and so of all the trillions of trillions of neutrinos passing through the Earth every second we would expect to capture a few of them with a large enough detector. Such detectors usually consist of tanks containing thousands of tons of dry-cleaning fluid. The chlorine in the dry-cleaning fluid is one of the best chemicals for interacting with the neutrinos. The tanks are placed deep underground so only neutrinos, which pass straight through the Earth, interact with the detector. Based on calculations of the rate of nuclear reactions in the Sun's core, experimenters predicted they would detect roughly one neutrino per day. Instead only one every three days was detected. This ratio of about one-third the predicted value has been confirmed by several different experiments and is known as the *solar neutrino problem*. The lack of detected neutrinos meant that either the scientists did not understand the nuclear fusion process as well as they thought they had, making all the calculations about the structure of the Sun wrong, or there was something about the neutrino that they did not understand. In the end, it was the scientists' understanding of neutrinos that was incomplete. Recently, scientists learned that neutrinos actually change type on their way from the Sun to the Earth. The chlorine detectors only detect the type of neutrino produced by the Sun's nuclear reactions—thus, because of the conversion en route, the number of detectable neutrinos is lower than the number produced by the Sun by about one-third. Raymond Davis and Masatoshi Koshiba received the Nobel Prize in 2002 for their experiments to detect neutrinos.

As an illustration of this process, imagine standing in a crowded gymnasium with each person holding an empty glass. There is a sink at one end of the gym and someone at the opposite end wants a drink, but because the gym is so crowded no one can move. The person nearest the sink can fill their glass with water and pour it into the glass next to them. This process could continue until the water is passed across the gym. This is similar to the Sun's energy being passed from atom to atom until it reaches the end of the radiation zone. One important difference between our analogy and what takes place in the Sun is that in the gym we would pass the water in such a way that it always moves towards the other side. The atoms in the Sun do not do this. In fact, there is no direct communication between one end of the radiation zone and the other. Instead, the Sun's energy is passed randomly from atom to atom. Sometimes it moves outward, sometimes inward, and just as often it moves side to side. Estimates vary, but generally it takes over 170,000 years for the energy released in the core of the Sun to get out of the radiation zone.

As the radiation travels through the radiation zone, the cooling temperature changes the conditions in the solar interior such that the transfer of energy through the absorption and reemission of radiation becomes less efficient, the atoms can still absorb the radiation well but do not release it as easily, and a new process takes over to transfer the energy the remaining 30% of the way to the surface. The outer layer of the solar interior is known as the *convection zone*, which gives a clue to the energy transfer process at work here. The most efficient means of transferring energy in this layer of the Sun is similar to the process by which you heat oatmeal or soup in a pot. The heat gets absorbed by the plasma at the base of the convection zone (top of radiation zone). This newly heated plasma rises, like the air in a hot air balloon, transferring the energy through the convection zone. As the rising plasma cools, it slows and begins to sink and fall back towards the base, only to be heated again to repeat its journey through the convection zone. This produces a rolling motion, with the energy being transferred in as little as a few days. Look at the top of a pot of boiling oatmeal or soup and you will see the same convective process at work there. The oatmeal is heated by the stove at the base of the pot and the hot soup rises to "pop" at the surface. A picture of the Sun shows that the solar surface is broken into patterns of cells, called *granulation*, which is caused by the rising and cooling plasma rolling over to sink back to the heat source.

Once the energy gets to the surface, the temperature there is such that there are no free electrons around, all of them having recombined with the atoms, so the energy that reaches the surface can escape unimpeded into space. The Sun is now shining.

THE LIFE OF THE SUN

The Sun is shining as a result of nuclear fusion converting hydrogen to helium in the core. This phase of the Sun is known as the *main sequence stage* and is where the Sun will spend the bulk of its active life ($\sim$90% of its life). Currently, the Sun is around 4.5 billion years old, with another 4.5 billion years of hydrogen burning to go. Once the supply of hydrogen in the core is exhausted, the star undergoes a sharp transition. The sharp falloff in the rate of nuclear reactions reduces the thermal pressure in the core and it can no longer support itself against gravity, so the core, which is now almost entirely made of helium, shrinks. As the core collapses, the increased temperature and density starts to reach the conditions in which the helium nuclei can now fuse, "burning" helium into carbon. The increasing rate of nuclear energy production produces so much energy that the outer layers of the star expand outwards against the gravitational contraction. The star grows in size by as much as 200 times, cooling as it does so. At the end of this expansion stage, the star is large enough to engulf the Earth and has cooled to the still toasty 3500 K. At this stage, the Sun will be known as a red giant star.

While this outer expansion is going on, the conversion of helium to carbon is progressing, producing the energy required to keep the star shining. Even though its temperature decreases, the much larger star requires more than 5,000 times the energy to sustain itself against the pull of gravity, and since the helium-carbon reaction is a less-efficient energy producer than the hydrogen-helium process, the helium-burning phase lasts only about 2 billion years. At the end of the helium-burning phase of a star like the Sun, no further nuclear reactions are possible and the Sun essentially ceases to shine. The last throes of helium-burning cause a sharp pulse of heat to travel through the outer atmosphere of the star, blowing off as much as 10% of the star's mass, which ultimately can form a planetary nebula, like the famous Dumbbell Nebula in Galaxy M27. It should be noted that a planetary nebula does not have anything to do with a planetary system. The name comes from the semblance of the nebula to the giant gas planets, like Jupiter and Saturn. The emission of the planetary nebula is a result of the interaction of the newly ejected mass from the star with mass ejected in earlier events. Once this interaction stops, there is nothing left of the star but a

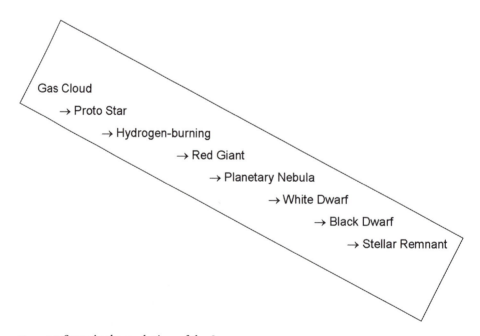

Figure 2.4 Steps in the evolution of the Sun.

slowly cooling remnant white dwarf about the size of the Earth but with a density some 200,000 times higher—a teaspoonful of matter from a white dwarf star would weigh about five metric tons. After billions of years of slowly radiating away like the embers of a campfire, the star will become invisible as a black dwarf. (More details about stellar evolution can be found in the *Stars* volume of this series.)

The basic stages in the Sun's life can be summarized in Figure 2.4.

While not particularly exciting as stars go, by virtue of being on our doorstep the Sun is preeminent in our drive to understand stars and the physical processes at work in the universe. Its impact on the Earth, our society, and technology notwithstanding, the Sun provides a unique glimpse into the inner workings of an astrophysical object. The following chapters take us deeper into our knowledge of the Sun and its impact on our understanding of the universe.

RECOMMENDED READING

Hansen, Carl J., Steven D. Kawaler, and Virginia Trimble. *Stellar Interiors—Physical Principles, Structure, and Evolution.* New York: Springer-Verlag, 2004.

Jones, Lauren V. *Stars.* Greenwood Guides to the Universe. Westport, CT: Greenwood Press, 2009.

Rivkin, Andrew. *Asteroids, Comets, and Dwarf Planets.* Greenwood Guides to the Universe. Westport, CT: Greenwood Press, 2009.

WEB SITES

Aurora: http://www.pfrr.alaska.edu/aurora.

Solar Neutrino Problem. See the article by John Bahcall at: http://nobelprize.org/nobel_prizes/physics/articles/bahcall.

Spectroscopy: http://loke.as.arizona.edu/~ckulesa/camp.

3

The Magnetic Sun

The Sun is a dynamically varying star exhibiting a wide range of complex behavior from the core to the surface and beyond. Through long and careful observation, a detailed picture of the Sun's complexities has emerged, and we now have a strong understanding of how the Sun transports energy from its generation in the nuclear fusion reactor at its center to its release as heat and light radiating into space. We know to very good accuracy the temperature and density structure in the solar interior, and are beginning to explore the dynamics of the interior through the science of helioseismology, which is allowing us to "see" inside the Sun (Chapter 4). Despite this depth of understanding, a number of challenges and puzzles remain. One of the most important, and most mysterious, of these is the Sun's magnetic field.

The heliosphere is pervaded by magnetic fields from the base of the convective zone inside the Sun to the very edge of the solar system. The magnetic nature of the Sun is one of its most important and, yet, most intriguing features. The magnetic field is the means by which a fraction of the energy, initially generated in the nuclear fusion reactions of the Sun's core, can be stored for future release in the solar atmosphere, driving most of the dramatic and energetic phenomena in the heliosphere—sunspots, solar flares, planetary magnetic storms. Despite its importance, we have only a rudimentary understanding of the most basic properties of solar magnetism. However, the Sun is a magnetic star and much of its interaction with the rest of the bodies of the solar system, and certainly the Earth, is governed by the form and behavior of its magnetic field.

In this chapter, we introduce the concept of magnetism, first from a historical perspective and then more directly as it relates to the Sun. The discussion in this chapter paves the way for subsequent chapters in which we discuss the wide array of magnetic phenomena that result from the complex physical system that is the Sun.

BRIEF HISTORY OF MAGNETISM

The phenomenon of magnetism has been known for thousands of years, with some of the earliest scientific studies dating back to the detailed observations of amber and lodestone by the Greek scientist, mathematician, and philosopher, Thales of Miletus (CA. 624–546 BC). Thales conducted a number of experiments observing that amber, when rubbed, attracted feathers and other light objects and that lodestone could attract iron. Lodestone is a naturally occurring iron oxide found around the world; lodestone is known as *magnetite* today. The word *magnetism* itself comes from the district around Miletus in Asia Minor (modern-day Turkey) known as Magnesia, where the lodestones were found. In the millennium following Thales' experiments, many theories were proposed to explain this mysterious attraction that seemed to invisibly spread across space, including Thales' own idea that the lodestones possessed a soul. It wasn't until the early seventeenth century that magnetism began to be truly understood.

The lack of understanding of why lodestones did what they did, however, did not prevent them from being extremely useful. In the latter part of the eleventh century, a tool came into use that was to have a profound effect on the world in the following six centuries: the *compass*. Widely thought to have originated in China as early as AD 200, the use of the compass ultimately paved the way for the successful navigation of the world's oceans, creating global powers such as Holland, France, Spain, Portugal, and Great Britain. Columbus's journey to the New World (1492) and Magellan's circumnavigation of the globe (1519–1522) could not have occurred without the compass.

In addition to revolutionizing the world through its impact on navigation, the compass also revolutionized the world of science. The fact that the compass needle always pointed to a single preferred direction (generally north) led to major scientific advances both in magnetism and in the general study of the Earth. By conducting a series of experiments with lodestone and a compass, the English doctor William Gilbert (1544–1603) discovered that the Earth itself was a giant magnet with both north and south poles. Gilbert discovered that the attractive force exhibited by rubbed amber was different from that displayed by the lodestone. He coined the phrase *electric*, the Greek word for amber, to describe this force and showed that other materials also exhibited an electric property when rubbed. Electricity and magnetism were being realized as distinct and separate forces. Gilbert also suggested that this newly discovered magnetic field of the Earth would also have a sphere of influence extending beyond the planet. Today we know this as the *magnetosphere*, a "magnetic bubble" about ten times the

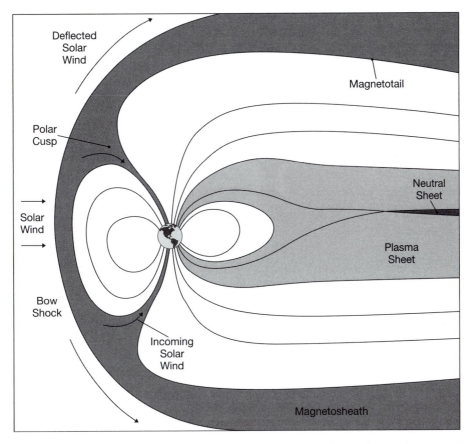

Figure 3.1 Schematic of the structure of the Earth's magnetosphere. The Sun is assumed to be to the left of the figure, with the solar wind blowing left to right. Illustration by Jeff Dixon.

size of the Earth, protecting the surface, and all life on the surface, from harmful radiation from the Sun (Figure 3.1). For reference, the magnetosphere was not actually discovered until the early days of the Space Age in the middle of the twentieth century, four hundred years after Gilbert's suggestion.

Before the advent of Global Positioning System (GPS) satellites for navigation, we relied on compasses for direction. The suspended needle (see sidebar) pointed north, and from there we could get our bearings and find the direction we needed. The discovery that the Earth had a magnetic field and that this magnetic field could be utilized for navigation was a major milestone in world history. Strictly speaking, however, the magnetized needle does not point to true geographic north, but to magnetic north. The Earth's magnetic field is similar in shape to that of a regular classroom bar magnet with a north and a south pole. The north magnetic pole is defined as the point on the Earth's surface where the magnetic field is directed vertically downwards; the south magnetic pole is where it is directed vertically

The Earth's Magnetosphere

Earth is surrounded by a magnetic field that originates deep in the Earth's core. The magnetosphere is the region of space that is dominated by this magnetic field and is shaped by the strength of the Earth's intrinsic magnetic field, the pressure of the solar wind, and the strength and direction of the interplanetary magnetic field. In the magnetosphere, magnetic and electric forces dominate those of gravity and control the movements of free electrons and ions that are generated as a result of the interaction between solar ultraviolet radiation and the Earth's atmosphere. The Earth's magnetosphere is not spherical, but actually has the shape of an eye with the anti-sunward side being elongated by the pressure of the solar wind to form a geomagnetic tail. The boundary of the magnetosphere is known as the *magnetopause* and is located about 15 Earth radii above the Earth on the sunward, or day, side with the geomagnetic tail extending out to about 200 Earth radii behind the Earth on the night side. The exact scale and shape of the magnetopause change with changes in the solar wind pressure. The magnetosphere provides a shield to fast-moving ions accelerated by shocks in the solar wind and solar flares by deflecting these particles to the north and south poles where they form the aurora. Without the magnetosphere, the solar wind would effectively sweep away the charged particles of the upper atmosphere, essentially generating a continuous erosion of the Earth's atmosphere.

upwards. This definition is opposite that normally applied to a simple bar magnet where the north pole is defined to be the pole where the field is directed *outwards*. By convention, and the fact that the inward-pointing magnetic pole lies near geographic north, this pole is known as the north magnetic pole.

How a Compass Works

A compass is a useful navigational tool that takes advantage of the fact that the Earth has a simple magnetic field with a single pair of magnetic poles, one positive (in the south) and the other negative (in the north). This is like the Earth having a large bar magnet under its surface with the south pole of the magnet lying under the geographic north pole of the Earth. Since like poles repel and opposite poles attract, this means that a small free-floating magnetic would align itself with its north pole pointing to geographic north.

To make a compass, take a small iron or nickel needle (iron and nickel are magnetic metals) and magnetize it by laying the eye of the needle on a magnet overnight. Once the needle is magnetized, it is necessary to find a way to allow it to turn freely. Commercial compasses center the needle on a narrow pivot point or suspend it by a thin string. A more convenient way at home is to push the needle through a small piece of cork and float the cork on a bowl of water. The magnetic forces from the Earth's magnetic field will cause it to spin so it is aligned in the opposite sense to the magnetic field of the Earth, that is, the needle will point to the two magnetic poles of the Earth, roughly along geographic north-south.

The angle measure between magnetic north and true geographic north is called the *magnetic declination*. The north magnetic pole lies in the

Canadian part of the arctic, while the south magnetic pole lies off the coast of Antarctica, south of Australia. Careful and frequent measurements have shown that the Earth's magnetic poles are not stationary, but slowly drift with time. In 2001, the north magnetic pole was located at latitude 81.3N longitude 110.8W, while in 2005 it had drifted to 82.7N and 114.4W. The pole is moving northwest at approximately 40 km per year. Following this progression, the pole should be approximately located at 84.0N and 118.1W on January 1, 2009. To monitor the constantly moving pole, frequent expeditions are required to pinpoint its exact location. In North America, this task is performed mostly by the Geological Survey of Canada, which carries out periodic surveys, the last of which began in 1998 and was completed in May 2001. In addition to the long-term drift, the pole also moves on a daily basis, sweeping out an elliptical path some 80 km about the average central location. The daily motion of the pole is due primarily to the interaction of the solar wind (see Chapter 8) with the Earth's magnetosphere. The charged particles that make up the solar wind generate electric currents in the Earth's magnetosphere and ionosphere, which, in turn, disturb the magnetic field. Fluctuations in the solar wind cause variable disturbances in the Earth's magnetic field, causing magnetic north to migrate throughout the day. The long-term variation, the northwest drift of the pole, is due to the dynamo process inside the Earth's outer core, 3000 km below the surface, which is responsible for generation of the field. This terrestrial dynamo is a consequence of the presence of an outer core of molten metal and the Earth's rotation. The combination of rotation and convection in the core generates circulating currents in the electrically conducting metallic core. This current creates and sustains a magnetic field that we measure at the surface and that extends into space as the magnetosphere.

The complete history of how we have come to understand magnetism would fill a volume on its own, so we can only touch upon it here. Over the 250 years following Gilbert's discovery, several other experiments were performed that would fully characterize the nature of magnetism and its forces. Parallel experiments were being performed with electricity, including American scientist Benjamin Franklin's kite-flying episode in which he discovered that lightning was a form of electrical spark. He went on to devise the lightning rod, which has been responsible for saving innumerable lives. The invention of the first-ever electrical battery by Italian scientist Alessandro Volta (1745–1827) in 1800 led to a remarkable discovery two years later when another Italian, Gian Domenica Romagnosi (1761–1835), noticed that the Volta battery caused the needle of a compass to deflect. This pointed to a direct connection between magnetism and electricity. This discovery went largely unnoticed until it was repeated by Danish physicist Hans Ørsted (1777–1851) in 1820. This time the scientific world paid attention and the field of electromagnetism was born. Over the succeeding decades, the observational and theoretical discoveries by scientists such as the French physicists André Marie Ampère (1775–1836) and Jean-Baptiste Biot

noted his observations of sunspots in 1610. Dutch student Johannes Fabricius and German astronomer Christopher Scheiner both published observations of sunspots in 1611 before Galileo's claim to have discovered them in 1613. Systematic sunspot observations since the mid-seventeenth century have proved extremely useful in our understanding of solar magnetism (this will be discussed more fully in Chapters 5 and 6). However, it has only been since the turn of the twentieth century that we have been able to unequivocally associate sunspots with regions of strong magnetic fields on the Sun. The fundamental importance of magnetic fields in the activity of the Sun has only been fully realized over the last one hundred years.

The discovery of solar magnetism came about when, in 1908, American astronomer George Ellery Hale (1868–1938) noticed that the fine linear structures emanating from sunspots bore a remarkable resemblance to the iron filing patterns formed around the poles of a standard bar magnet (see sidebar). To test this idea, Hale used a solar spectrograph, which he had invented four years earlier at the recently completed solar tower telescope at Mount Wilson near Pasadena, California, to look for signs of a special splitting of solar absorption lines (see Chapter 2) that could only be caused by the presence of a strong magnetic field. This is known as Zeeman splitting, named after Dutch physicist Pieter Zeeman (1865–1943) who discovered this effect in 1896 (see sidebar). Hale was immediately rewarded by the observation of a clear line triplet indicating that sunspots contained strong magnetic fields. This knowledge led to fundamental advances in solar physics; Hale's legacy continues today with magnetic field measurements of sunspots and surrounding areas driving much of today's solar physics research.

Zeeman Splitting

Spectral lines are produced when electrons gain or lose energy within an atom (see Chapter 2). An atom of a given element has a distinct set of energies available to the electrons that surround the nucleus. When an electron changes from one energy to another, a line at a distinct wavelength is observed—as a bright emission line if the electron loses energy or a dark absorption line if the electron gains energy. However, Dutch physicist Pieter Zeeman found that for atoms placed in a uniform magnetic field, the single spectral line is split into a set of closely spaced lines comprised of the original line accompanied by two additional lines symmetrically placed: one at a slightly longer wavelength and the other at a slightly shorter wavelength. This splitting of the line is an effect of quantum mechanics and is associated with what is called the orbital angular momentum quantum number, L, of the atomic energy level. The magnetic field causes a wobble in the atom's spin, resulting in a split of the energy levels. For a given value of the quantum number, L, the resulting number of split levels is 2L + 1: L = 1 refers to the first excited state of the atom and so commonly the strongest Zeeman effect is a triplet line centered on the corresponding unmagnetized wavelength. The size of the splitting of the lines depends directly upon the strength of the magnetic field. Thus, a detection of the Zeeman line splitting indicates the presence of a magnetic field, and the size of the splitting determines the strength of the magnetic field.

Hale's discovery of magnetic fields in sunspots and the series of observations that followed were limited to regions of the Sun containing strong magnetic fields. This was because of the limited ability of the spectrographs of the time to resolve small line splittings. Significant advances in the sensitivity of solar spectrographs followed Hale's initial discovery, and the sensitivity to the Zeeman effect on optical lines with wavelengths around 630 nm improved considerably. Ultimately, magnetic field strengths as small as 200 G could be measured, but this still restricted the observations to the regions around sunspots. This all changed with the invention of the solar magnetograph by American father-and-son team Harold and Horace Babcock. In the aftermath of World War II, the new technologies of photomultipliers and solid-state electro-optic modulators became readily available, and Horace Babcock developed the photoelectric magnetograph. In 1952, the Babcocks used their magnetograph to discover that the whole surface of the Sun was literally covered with magnetic field. The solar magnetograph introduced an improvement of about two orders of magnitude in the sensitivity of magnetic field measurements over the contemporary visual or photographic techniques developed by Hale and collaborators. Without these observational breakthroughs, we would know very little of the importance of the magnetic field for activity on the Sun. The solar magnetograph is such an important tool for astronomers that modern versions of this instrument are used in virtually every major solar observatory in the world; there are even, at present, two magnetographs on spacecraft observing the Sun.

THE MAGNETIC CARPET

Today the solar magnetic field is routinely monitored, observed continuously from an array of ground-based observatories around the world and also from an increasing number of observatories in space. Not only can we now measure the strength of the magnetic field, but also the direction in which the lines of force point. This is important information as it enables solar astronomers to determine how much energy is present in the field and how much can potentially be released in a solar storm. The variation of this field, its strength and direction, tells us a lot about the energy processes that power the Sun's atmosphere (see Chapter 7).

As the observations of the Babcocks demonstrated, evidence for the presence of magnetic field is found wherever you look on the Sun. The strong field is contained in sunspots, but these occupy only a small fraction of the solar surface at any one time. An understanding of the Sun's behavior would not be complete without an understanding of the less-intense and smaller-scale fields distributed widely around the solar photosphere. At any one time, magnetic measurements show that the Sun is covered by tens of thousands of small magnetic bipoles—a pair of north and south pole magnetic fields connected by magnetic lines of force that protrude out of the photosphere. When measuring magnetic fields, solar astronomers typically use black to represent negative, or south, poles and white to indicate positive, or north, poles, and so a magnetic picture of the Sun shows a salt-and-pepper smattering of magnetic bipoles. Closer inspection of this salt-and-pepper pattern shows that the magnetic concentrations are not randomly located. The observed magnetic field is found to be concentrated on the boundaries of large cells associated with the granulation caused by solar convection (see Chapters 2 and 5). This small-scale magnetic field, known as the quiet network field, is very dynamic and is connected to the larger fields of the sunspots. Recent analysis of the quiet solar network field, using data from the *Solar and Heliospheric Observatory* (SOHO) spacecraft, has shown that it is recycled on a timescale of eight to nineteen hours. In other words, the quiet network is constantly changing (colliding, disappearing, re-appearing) with the whole network being replenished, on average, every twelve hours. This recent discovery can be contrasted with conclusions reached in the 1940s that the average lifetime of solar magnetic field should be of the order of centuries; the plasma that makes up the solar atmosphere is such a good conductor that the field decays very slowly.

Decay of Magnetic Field

The solar atmosphere is almost a perfect electrical conductor, but not quite. The connection between electric and magnetic fields discussed earlier in this chapter indicates that changing magnetic fields can create electrical currents and vice versa. In a non-perfect conductor, some of the energy in the currents flowing can be dissipated as heat and light, which is essentially how incandescent light bulbs work. In this way, energy can be removed from the magnetic field. In the solar atmosphere, a magnetic field, if left alone, will decay on a timescale defined by the properties of the plasma making up the solar atmosphere. For the solar corona, a field of about 100 Gauss (10^{-2} Tesla) will decay away in about one million years.

The rapid recycling of the many small magnetic concentrations that make up the quiet network, commonly called the *magnetic carpet*, suggests that there is a continuous process whereby magnetic energy is transferred from the photosphere and below into the solar atmosphere. Such a continuous supply of energy is required to maintain the million-degree temperature of the solar corona (see Chapter 7). The presence of the magnetic carpet also implies that the quiet network field must be generated locally and that very little of it is due to magnetic field spreading out from sunspots. In other words, there are two independent sources of magnetic field in the Sun, the large-scale production of field that produces sunspots and the global magnetic field that surrounds the Sun, varying over a timescale of several years, and the rapidly changing small-scale field that makes up the magnetic carpet, varying over a timescale of a several hours. These observations provide insight into the generation of magnetic field on the

Sun known as the solar dynamo. The large-scale solar magnetic field is thought to be a result of dynamo action deep in the solar interior, while the presence of a magnetic carpet suggests that some form of dynamo action is also occurring just below the solar surface, perhaps caused by swirling turbulent motions that exist at the top of the convection zone.

THE SOLAR DYNAMO

In the previous sections, we discussed the properties of the Sun's magnetic field, its first observation, its division into sunspot and network fields, and its interaction with the solar atmosphere. An important question to ask here is: where did the magnetic field come from and how is it maintained? The answer lies in the interior of the Sun and the physics of the interaction of magnetic and electric fields. In the earliest experiments using magnetism and electricity, it was found that passing an electric current through a conducting wire generated a magnetic field—detected by a deflection of a compass needle. This discovery that electric currents could generate magnetic fields was made in 1820 by Hans Ørsted. Eleven years later, in 1831, Michael Faraday, the discoverer of magnetic lines of force, demonstrated that a magnetic field moving through a coil of conducting wire (copper) generated a small electric current that flowed through the wire. Thus, moving electric charge, that is, an electrical current, can generate magnetic fields, while moving magnetic fields can generate electric currents.

In the interior of the Sun, all of the necessary elements to generate magnetic field are available. While there are no metal wires, the high temperature of the solar interior causes most of the atoms in the gas (predominantly hydrogen atoms) to be ionized, that is, the intense heat causes the electrons to separate from the atoms, leaving charged nuclei and free electrons. Such a gas is called *a plasma*. In a plasma, the electrons are then free to move around, much like the electrons in a metal wire. The other piece of the puzzle is provided by the fact that the Sun's interior is constantly moving. Two of the main forms of motion of the plasma in the interior of the Sun are rotation and convection (see Chapter 2). These motions cause the electrons to flow around the interior, generating a magnetic field. Only certain types of flow, however, allow the field to grow as it draws energy from the flows. The rotational flows on the Sun are quite complicated. The inner 70% of the Sun (the core and radiation zone) rotates essentially like a solid ball with all parts sweeping out the same angle in the same time. The outer 30% (the convection zone and surface), however, rotates differentially with the plasma at different latitudes rotating at different rates, the equator moving faster than the poles (see sidebar). This mismatch between the flows in the convection zone and those of the radiation zone creates a layer, known as a shear layer, or *tachocline* (see next chapter), where the magnetic field is magnified. The strengthening magnetic field becomes buoyant due to the growing magnetic pressure and works its way up to the surface to appear as a pair of sunspots (see Chapter 4 for more detail).

Differential Rotation

It has been known since the first systematic observations of sunspots in the early 1600s that the Sun rotates. Sunspots are seen to march across the face of the Sun at a regular rate, indicating that the Sun rotates once on its axis, on average, approximately every twenty-seven days. This is equivalent to the Sun rotating at roughly 6,800 km/hr. More careful observations show that sunspots at different solar latitudes rotate around the Sun at different rates, with the speed getting slower the further the sunspot is from the equator. Thus, the higher the latitude of the sunspot (north or south) the more slowly it is observed to move around the Sun. Regions near the solar equator take about twenty-six days to go all the way around, while regions near the pole take more than thirty days. The Sun is, therefore, said to be rotating differentially with different latitudes rotating at different rates. This turns out to be quite important for the process of generating magnetic fields.

The generation of the Sun's magnetic field is quite a complex issue, as any successful model must not only explain the production of sunspots, but also the various patterns of behavior they exhibit, all of which will be discussed in detail in the chapters that follow. These patterns include:

- Solar Cycle: The average number of sunspots visible on the solar surface, the number of solar flares that occur, and a number of other phenomena, exhibit an eleven-year repeating variation.
- Spörer's Law: An equator-ward progression of the latitude at which sunspots appear as the cycle progresses, named after German astronomer Gustav Spörer (1822–1895), and best illustrated by the *butterfly diagram*, indicating the latitudinal variation of sunspot emergence over the course of a solar cycle.
- Hale-Nicholson Polarity Law: Sunspot pairs are observed to have a leading and a following polarity, with the leading sunspot traveling around the Sun ahead of the following polarity. Hale's law notes that in a given solar cycle each leading spot in the northern hemisphere has the same polarity, but that this polarity is opposite that of the leading spots in the southern hemisphere, for example, if all leading spots in the northern hemisphere are north polarity, then all leading spots in the southern hemisphere are south polarity and vice versa. The polarity of the leading spot changes with each solar cycle.
- Joy's Law: All sunspot groups tilt towards the equator, that is, regardless of hemisphere and cycle, the leading spot is closer to the equator than the following spot.
- The magnetic fields at the physical poles of the Sun reverse polarity at the time of solar maximum, that is, the time when the Sun has its maximum number of sunspots.

These are fundamental properties of the Sun that all point to a specific behavior of the magnetic field and the processes that generate it. All must be explained by any credible model of the solar dynamo.

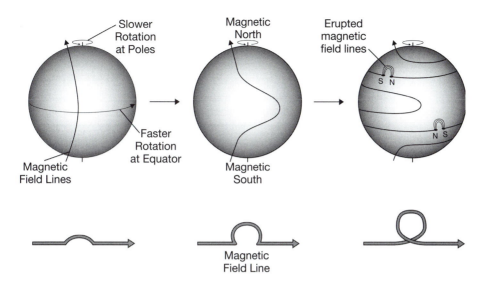

Figure 3.3 Solar dynamo alpha-omega effect showing the conversion of poloidal field to toroidal field. Illustration by Jeff Dixon.

Generally, the basic properties of the solar dynamo can be described by the combination of two effects: the omega (Ω) effect and the alpha (α) effect. As a starting point, we begin with a field line that runs directly between the north and south poles of the Sun (see Figure 3.3). This is known as a poloidal field line. As the Sun rotates, this field line gets dragged around, but because the Sun rotates differentially with the rate getting progressively slower from the equator to the poles, the field at the equator gets dragged around faster and the field is stretched in the longitudinal (or toroidal) direction. This is the Ω-effect, named after the Greek symbol Ω, used to represent the Sun's rotational velocity. In this way, a poloidal field is converted into a toroidal field. If this was all that happened, then all of the poloidal field would be converted over time into a toroidal field, that is, a one-way path, and the cyclical behavior of the Sun could not be explained.

To complete the dynamo process, it is necessary to allow the toroidal field to convert back to a poloidal field so the process can start all over again. This is assumed to occur via the convection flows that make up the outer 30% of the Sun's interior. A toroidal field line that gets picked up by a convective cell will be pulled upwards and twisted such that the field line gets reoriented from the toroidal to the poloidal direction.

The twist of the rising field line is caused by the *Coriolis effect* (see sidebar) that results from the Sun's rotation. This twisting is known as the α-effect. The poloidal field generated this way points in the opposite direction from the original poloidal field, satisfying the field-reversal property of the Sun's magnetic behavior. This poloidal field then gets stretched out by the solar rotation, and the process starts all over again.

The Coriolis Effect

The Coriolis effect was first used to describe the deviation of winds on the Earth from straight paths and was named after the French scientist Gustave-Gaspard Coriolis (1792–1843), who described it in 1835. Differences in pressure in the atmosphere tend to drive the air from the higher to lower pressure regions. As the air flows, the Earth rotates under it, making the wind flow in a curved path. Because the Earth is a sphere that rotates at a fixed rate of 360 degrees every twenty-four hours, a point on the equator has to travel a longer distance in that one day than a point at any other latitude—the circular paths at higher latitudes are smaller, but still take twenty-four hours to complete one turn. For example, a point on the equator is moving faster than a point on the Arctic Circle. Air moving northwards from the tropics has extra speed relative to the air it is running into and so deflects to the right in the direction of rotation, that is, eastwards. Similarly, in the southern hemisphere, air moving from the tropics towards the Antarctic Circle deflects to the left, also in the direction of the Earth's rotation, that is, eastwards. This explains why large storms circulate counterclockwise in the northern hemisphere and clockwise in the southern hemisphere. In the Sun, magnetic fields rising through the convection zone get deflected by the rotating plasma, causing a change in the direction of the magnetic field, producing toroidal field from poloidal field.

THE MAGNETIC SUN

The discovery of the Sun's magnetic nature by Hale in 1908 and the invention of the magnetograph by the Babcocks in 1953 are pivotal points in the history of solar physics. We now know that the Sun's atmosphere is filled with magnetic field and that this magnetic field is responsible for a wide range of solar activity, from the formation of sunspots, to the heating of the atmosphere to millions of degrees, to the production of energetic solar storms. Even the quiet Sun, away from sunspots, displays a startling level of activity and dynamics. All of this diverse activity is magnetic in origin. The key to understanding the activity of the Sun, then, is to better understand its ever-changing magnetic field. We will explore this range of activity in the remainder of this volume.

WEB SITES

Build Your Own Magnetometer in a Bottle: http://image.gsfc.nasa.gov/poetry/workbook/magnet.html.

Earth's Magnetic Field: http://www-istp.gsfc.nasa.gov/earthmag.

History of Magnetism: http://www.aacg.bham.ac.uk/magnetic_materials/history.htm.

History of Navigation: http://www.abc.net.au/navigators/navigation/history.htm.

Magnetic Carpet: http://www.lmsal.com/magnetic.htm.

SOHO: http://sohowww.nascom.nasa.gov.

Sun's Magnetic Field: http://solarscience.msfc.nasa.gov/dynamo.shtml.

4

Looking Inside the Sun

In 1962, a historic observation was made that changed the way we viewed the Sun and opened the way for a brand new area of research called *helio-seismology*. As its name implies, helioseismology is the study of the seismology of the Sun—studying how the Sun shakes. Robert Leighton, Robert Noyes, and George Simon, solar physicists at Caltech in Pasadena, California, identified an interesting pattern in their observations of the solar surface: the Sun seemed to be ringing like a bell! Subsequent observations by Leighton and his colleagues demonstrated that everywhere one looked on the Sun, the signal varied in a regular oscillatory pattern with a distinctive period of five minutes. These five-minute oscillations, where the surface rises and falls in a steady rhythm like the breathing of a sleeping child, were described by Roger Ulrich, also of Caltech, as the result of sound waves that travel through the interior of the Sun, occasionally sticking their heads above the surface. The oscillations can be as high as tens of kilometers and travel at several hundred meters per second. A ten-kilometer wave may sound like something to avoid, but the 700,000 km radius of the Sun makes this wave seem rather small and consequently difficult to detect. The regular five-minute pulse of these solar sound waves corresponds to a frequency of about 3 mHz or 0.003 vibrations every second. Even if these waves could travel across the vacuum of space to the Earth, their frequency is about 10,000 times lower than a human ear can register.

Modern study has found that the Sun rings at more than one hundred thousand different frequencies, each sampling a different tone of sound wave traveling around the Sun. The regularity of these oscillating patterns indicates that the sound waves are trapped inside the Sun and are

43

continuously produced. The different frequencies indicate waves that reach different depths into the solar interior. By analyzing all of the observed frequencies using modern computers, models, and mathematical techniques, solar scientists can probe in detail the interior structure of the Sun. This discovery and the subsequent theoretical explanation allowed the interior of the Sun to be investigated with unprecedented precision. Forty-seven years later, solar astronomers now routinely monitor the five-minute oscillations and the tens of thousands of other frequencies that make up the Sun's oscillatory behavior. Not only do we have a better understanding of the solar interior, but we are also now able to detect "events," such as the emergence of a new sunspot, on the far side of the Sun. Based on these observations, sophisticated models of the solar interior have been developed that are so accurate they led to the solution of the solar neutrino problem, making a contribution to fundamental physics by proving categorically that neutrinos have mass. This science of helioseismology has led to major breakthroughs in our understanding of the structure of stars, their internal energy distribution, and their ability to generate magnetic field. In this chapter, we will explore the ringing of the Sun and discuss what we have learned since that first discovery more than forty-five years ago.

Wave Motions

We are all familiar with many different waves that we see, but perhaps not notice, every single day. The ripples caused by a rock thrown into a pond, the breakers that crash onto the beach, the sound of your favorite band emanating from your headphones, or the voice of a friend calling you, are all examples of wave motion in action. The passage of a wave through a medium like water or air is a traveling disturbance in which the particles of the medium are caused to oscillate back and forth. For solar oscillations, we are also interested in the physics of sound waves. Sound waves are known as longitudinal waves, where the particles of the medium are displaced in the same direction as the wave propagates. The particles themselves do not move with the wave, but merely oscillate back and forth. A tuning fork creates a sound wave that travels by alternately compressing and stretching the air it passes through, creating a traveling pattern of pressure changes that is picked up by the mechanics of your ear. Each particle pushes on its neighboring particle, causing it to push it forward and transmit the wave.

Waves are generally described by three interrelated properties: their wavelength, λ, their frequency, v, and their speed, c_s. The wavelength describes the separation between successive peaks (or troughs) of the wave, the frequency describes how many peaks, or troughs, pass a given point every second, and the speed describes how fast the wave is propagating. These quantities are connected via the relationship:

$$\lambda \times v = c_s.$$

An additional quantity used to define a wave is its amplitude. This is effectively the maximum height, or displacement, of the wave perpendicular to its direction of travel—the energy being transported by the wave is proportional to the square of the amplitude.

A special case occurs when the medium through which the wave is passing is fixed at both ends, or wraps around on itself. For example, plucking a string of a guitar produces a sound of a specific frequency, determined by the length of the string. Since the length of the string is fixed, the waves generated travel up and down the string reflecting back off the fixed ends. Only those waves whose wavelength is a multiple or an integer fraction of the length survives, creating a standing wave of a given wavelength (and therefore a pure note). The points on the string that do not move up or down, but are fixed, are called nodes; those points that have the maximum amplitude are called antinodes. Strictly speaking, a standing wave is not a wave but a pattern of nodes and antinodes created by the interference of two waves, the original wave and the reflected wave. Two obvious nodes are the two ends of a fixed string that by definition have zero displacement. The fundamental frequency of the string is then defined directly by the length of the string with a single node at either end and no other nodes anywhere along its length. For this frequency the string rises and falls as one. Harmonics of this fundamental frequency occur when one or more additional nodes are present along the length of the string. To maintain the standing wave pattern, the successive nodes have to be located at integer fractions of the string length. Thus, the second harmonic has a node half way along the string, while the third harmonic has two additional nodes, located a third of the length of the string in from each end, and so on. It is the creation of a standing wave pattern of nodes and antinodes that defines the oscillatory nature of the Sun where effectively thousands of strings are being plucked at once, each having many harmonics.

SOLAR OSCILLATIONS

The discovery of the five-minute oscillations by Leighton and colleagues was a major milestone in our developing understanding of the Sun as a star. In the decade following the initial observations in 1962, it was widely thought that the oscillations were a surface phenomenon, with Leighton stating that the phenomenon was a consequence of the "local properties of the solar atmosphere." However, the seminal paper of Roger Ulrich in 1970 turned that on its head, and the true origin of the oscillations was finally explained. Ulrich characterized the oscillations as standing sound waves that travel around the Sun trapped in a narrow layer below the solar surface (Figure 4.1). At the top end of this layer, the waves reflect off the solar surface because of the sharp density decrease between the photosphere (surface) and the overlying solar atmosphere. At the bottom end, the increasing temperature causes the waves to refract (or bend), much like light passing through a prism, eventually turning completely around and returning to the surface only to be reflected again, and so on all around the Sun. The penetration depth of a given wave depends on the angle it makes to the surface, which in turn depends on the frequency and the wavelength along the surface. Waves with small wavelengths are confined to the topmost layers of the solar interior, while large wavelength waves can penetrate deeper into the Sun. When the wavelengths are such that after a complete passage around the Sun the wave arrives at its exact starting point, we have a standing wave, and subsequent passages of the sound wave reinforce the size of the oscillations making them detectable. This was subsequently and independently confirmed by other theorists.

The reason Leighton and colleagues had not detected the global nature of these waves was because of the short time series and limited spatial and

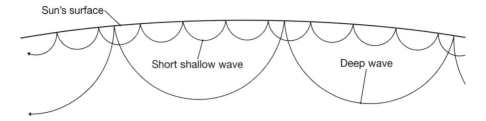

Sun's surface

Short shallow wave

Deep wave

Figure 4.1 Standing waves on the Sun. Waves of different modes penetrate to different depths in the Sun. Illustration by Jeff Dixon.

frequency resolution of their observations. As a result they were unable to pick out a single coherent frequency among the jumble of waves that comprise the oscillatory motion of the Sun. After the explanation provided by Ulrich, improved observations were made to test the predictions of the standing wave theory. The exact details of the theory indicate that the specific frequencies of the waves are determined by the internal structure and dynamics of the layer through which the waves travel. The standing wave model predicted a very precise relationship between the wavelengths of the waves and their frequency, whereby the strongest oscillations were found to lie on well-defined ridges. Detailed observations found a pattern that matched perfectly the pattern predicted by the models.

Because the Sun is almost perfectly spherical, the different modes of oscillation of the Sun can be described by three distinct integer numbers, usually denoted l, m, and n, with l and m representing the horizontal oscillations and n representing the oscillatory motion in the radial direction. The oscillation behavior of the Sun at any given frequency is made up of a combination of these different modes (see Figure 4.2). The numbers can be chosen independently, but two constraints that must be satisfied are:

(a) $l \geq 0$, and
(b) the integer m must take a value in the range $[-l, +l]$.

In other words, m can be positive or negative, but cannot have a magnitude greater than l. Negative values of m correspond to waves that travel opposite to the direction of the solar rotation, while positive values travel with the sense of rotation. The integer n can take any integral value, positive or negative (Figure 4.2).

For a basic one-dimensional standing wave (see sidebar), the nodes are defined to be those being in the negative (or positive) direction, that is, where the amplitude of the wave is zero. In the more complicated three-dimensional oscillation of the Sun, these locations are defined not by points but by circles around the Sun—nodal lines. The values of the integers l, m, and n define the number of nodal circles that specify a particular oscillation. The integer l denotes the total number of nodal lines on the surface, regardless of whether they lie in the latitudinal or longitudinal direction, while the

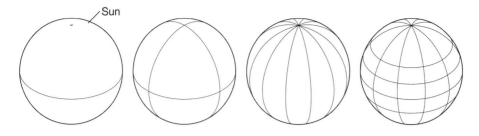

Figure 4.2 Schematic of sample global modes of oscillation of the Sun. Illustration by Jeff Dixon.

integer m denotes the number of nodal lines that cross the equator. This explains why $|m| \leq |l|$.

A full description of all the possible modes of solar oscillation would take an entire book in itself. For illustrative purposes we focus on the angular degree integer, l, as this defines the spatial scale of the waves. The larger l is, the more nodal circles are present, and, therefore, the smaller the wavelength of the sound wave responsible for the oscillation and the shallower the penetration of the wave into the interior. The smaller l is, the larger the wavelength and the deeper the wave penetrates (Figure 4.1). The $l = 0$ mode is called the "breathing" mode where the whole solar surface moves in and out at the same time. The combination of all the modes and an understanding of how far they penetrate into the Sun tell us much about the medium the waves pass through and therefore much about the interior structure of the Sun.

Observationally, the trick is to observe the Sun continuously for as long as possible; the study of solar oscillations requires long, nearly continuous observations. In this way, the many different wavelengths of the waves traveling around the Sun can be "picked out." Obviously, this cannot be done by a single observatory on the Earth as the rotation of the Earth results in a "nighttime" gap of more than twelve hours depending upon the location and time of year. To compensate for this, two different approaches have been taken. Several helioseismology networks of ground-based observatories exist, distributed at different longitudes to provide continuous observation of the Sun. Each observing station in the network has the same kind of telescope running the same way. In this manner, a complete and consistent set of solar observations can be obtained. Data from each station are combined into one contiguous data set for careful analysis. One of the most prominent of these networks is the Global Oscillation Network Group (GONG), which spans the globe with six ground stations. GONG has been operating since 1995, and the frequencies of almost half-a-million oscillation modes have been calculated. Each of these modes provides information about a different part of the solar interior.

In addition to ground-based telescopes, helioseismology observations have been taken from space, where a telescope in the correct orbit can have a continuous unobstructed view of the Sun. One such orbit is that per-formed by the Solar and Heliospheric Observatory (SOHO) satellite. SOHO orbits a special gravitational balance point known as the L1 Lagran-gian point. This is a location in space roughly one-and-a-half million kilo-meters closer to the Sun, along the Sun-Earth line, where the gravitational pull of the Sun and the Earth are equal. SOHO was launched in December 1995 with a suite of twelve instruments that have produced the most re-markable observations of every aspect of the Sun from the interior to the solar wind. One of these instruments, the Solar Oscillations Investigation Michelson Doppler Imager (SOI/MDI), is designed to observe solar oscilla-tions at high spatial resolution. The thirteen years of data from SOI/MDI and GONG have revolutionized our view of the interior structure of the Sun. The oscillation modes of the Sun represent the addition at each point on the Sun of millions of standing waves with each mode having an ampli-tude of only a few centimeters per second. Observations are now so good that each mode of oscillation of the Sun can be determined to an accuracy of less than one part in ten thousand.

The quality of these measurements has allowed solar scientists to accu-rately determine the speed of sound and how it varies within the interior, the density structure of the inside of our star, how fast the inside of the Sun is rotating, where the convection zone starts (see Chapter 2), and even the correct radius of the Sun. The measurements have also contributed to the development of better and more sophisticated theories of stars, which can then be used to understand astrophysical processes elsewhere in the universe.

PROBING THE SOLAR INTERIOR

As observations have improved, our detailed knowledge of the oscillation modes and what drives them has led to a much greater understanding of the structure and dynamics of the solar interior. As the thousands of differ-ent waves pass through the different regions of the solar interior, they act like probes, providing information on densities, temperatures, sound speeds, and motions. When these data are combined with advanced com-puter modeling, the solar interior has become one of the most understood regions in astrophysics.

The science of helioseismology has had many successes, particularly in the last twelve years since the implementation of the GONG (see sidebar) and the launch of the SOI/MDI instrumentation on board SOHO. These include the discovery that the standard solar model, used to describe the in-ternal workings of the Sun as a star, contained significant errors with respect to the composition, rotation, and even scale. Helioseismic observations led to a major improvement of the model and, therefore, our understanding of the Sun:

Helioseismology Observations

Since the first observations of solar oscillations, many observatories have undertaken long-term studies of the various modes of oscillation, their wavelengths, and frequencies. To separate the throng of waves, one must observe the Sun as continuously as possible for a long time. The longer one observes, the more readily the periodicities can be extracted.

Modern helioseismology utilizes networks of ground stations, where the Sun is always observed by at least one member of the network, and observations from space, where nighttime is not an issue. Table 4.1 summarizes some of the more prominent helioseismology networks and observatories operating at present.

Table 4.1. **Active helioseismology experiments**

NETWORK/OBSERVATORY	ACRONYM	# OF STATIONS	MODE RANGE
Ground-based			
Birmingham Solar Oscillations Network	BiSON	6	$l \leq 2$
Global Oscillations Network Group	GONG	6	$l < 250$
High Degree Helioseismology Network	HiDHN	2	$l < 500$
International Research on the Interior of the Sun	IRIS	8	$l \leq 7$
Low-L	LOWL	1	$l < 100$
Taiwan Oscillation Network	TON	5	$l < 1000$
Space-based			
Solar Optical Telescope[1]	SOT	–	
Global Oscillations at Low Frequencies[2]	GOLF	–	$l \leq 2$
Solar Oscillations Investigation[2]	SOI/MDI	–	$l < 1500$
Variability of solar IRradiance and Gravity Oscillations[2]	VIRGO	–	$l \leq 7$

[1] On board Hinode spacecraft.
[2] On board SOHO spacecraft.

- Following Roger Ulrich's original explanation of the five-minute oscillations, calculations by University of Cambridge scientist Douglas Gough demonstrated that the convection zone extended deeper into the interior of the Sun than previously thought.
- The analysis of the solar oscillations has yielded the detailed characteristics of the speed of sound throughout the solar interior. The sound speed is directly related to the temperature of the plasma, and as such provides important information on how energy is transported throughout the convection zone.
- Helioseismic studies of the internal rotation of the Sun showed that previous ideas on how the Sun generated magnetic field were wrong, and new models

of the solar dynamo resulted (see below). Of particular note was the discovery of a narrow shear region called the tachocline (see sidebar), which lies at the base of the convection zone. The tachocline separates the rotation of the differential rotation of the convection zone from the rigid rotation of the radiation zone.

- Analysis of oscillation modes that sampled the solar core confirmed the calculations of the rate of production of neutrinos in the Sun's hydrogen burning core, helping to identify the solar neutrino problem (see below) as a problem with our understanding of fundamental physics rather than of the solar interior.

- The discovery of a plasma "jet stream" about 20,000 km below the solar surface may shed some light on the generation of magnetic field and the regulation of the eleven-year solar cycle. This jet stream, similar in form to the jet stream that affects U.S. weather patterns on Earth, consists of flat egg-shaped regions some 30,000 km across traveling about 130 km/hr faster than their surroundings.

The Tachocline

At the base of the convection zone, just before the transition to the radiative zone, there is a narrow region known as the *tachocline* (from the Greek *tachos*—ταχοξ—meaning speed and *cline*—κλινοξ—meaning turn) where the rotation of the Sun makes a remarkable transition. Above the tachocline, the convection zone rotates differentially, that is, different latitudes rotate at different rates. Below the tachocline, the radiative zone rotates rigidly, that is, the whole zone rotates at the same rate like a spinning baseball. The transition between the two distinct rotational behaviors is very sharp, with the tachocline being not more than a few percent of the total solar radius, or about 28,000 km.

The discovery of the tachocline led to important modifications to our understanding of the physics of the solar interior, most importantly to the rotational characteristics of the Sun and to the theories of how the Sun generates magnetic field.

As mentioned previously, a major result of the observations of the global modes of solar oscillations has been the determination of the solar rotation rate as a function of location throughout the solar interior. The determination of the rotation is a complicated process that combines the information from all of the different frequencies to infer the physical characteristics of the plasma through which the waves travel. The details are too complex to discuss here, but a key feature is the effect that the actual rotation has on the oscillation frequencies themselves. In the previous chapter we discussed how the presence of a magnetic field can cause a split in the emission lines and that the size of that split allows one to infer the magnetic field. This was known as Zeeman splitting. When dealing with solar oscillations, where the waves travel all around the Sun, the wave frequencies are split depending upon whether the wave is traveling in the direction of solar rotation (east to west) or against it. Waves that travel with the solar rotation would appear to move faster than waves traveling opposite to the rotation (think of walking up a down-escalator compared to walking up an up-escalator). When observed from the Earth, the faster waves would register as waves with a slightly higher frequency, while the slower waves would have a slightly lower frequency than an identical wave in a nonrotating Sun. This frequency-splitting is directly dependent on the speed of the Sun's rotation. By looking for the frequency splitting on different waves, which penetrate to different depths, a complete profile of the internal rotation of the Sun can be deduced.

Figure 4.3 illustrates the distribution of the rotation speed in the solar interior. In the convection zone the speed of rotation depends on latitude, with the equator rotating faster than the poles—this is known as differential rotation. Thus, at the equator the whole convection zone rotates once every twenty-five days or so, while at the poles the rotation rate drops to about thirty days. This behavior is similar to that observed on the surface using sunspots and other surface features. The important consequence of these observations, however, is that there is no variation in the rotation speed with depth into the convection zone. Early models of solar rotation predicted that there would be a significant variation with depth. New models are still trying to fully incorporate the helioseismological observations and their implications.

In the radiative interior the rotation curves all converge to roughly a single value regardless of latitude or depth—the radiative interior is rotating rigidly as a single spherical body. To match up the rigid rotation of the

Figure 4.3 Plot of solar rotation as a function of depth in the solar interior showing transition from differential rotation in the convection zone to almost rigid rotation in the radiative zone. Courtesy of NASA.

radiative zone and the latitude-dependent rotation of the convection zone, a transition layer is required (the tachocline). The tachocline is the region where the convection zone speeds at low latitudes slow down to match the rigid speed of the radiative zone, while the convection zone speeds at high latitudes speed up. It is now generally held that the interesting dynamics of the tachocline are responsible for generating the Sun's large-scale magnetic field (see Chapter 3) and the observed solar cycle.

LOCAL HELIOSEISMOLOGY

Up until now, we have been discussing oscillations of the whole Sun, standing waves that resonate all around the solar surface. These are known as global modes, and the analysis of these modes is called global helioseismology. However, to investigate more local effects, like the appearance of a sunspot, new techniques under the title of *local helioseismology* have been developed. In these techniques, measurements of wave travel time between different points on the solar surface are used to provide a three-dimensional view of the solar interior and to detect localized structures in temperature, density, and magnetic field. For example, because the sound speed is dependent on temperature, any hot spot along the path of a wave will change its speed and result in a change in the travel time. The same applies to regions of higher or lower density or magnetic field.

The field of local helioseismology has been developed over the last several years, and as the techniques improve important new discoveries are being made. In particular, several new and significant flow patterns have been revealed within the shear layer that lies at the top of the convection zone. These include a jet stream at high latitude ($\sim75°$) with accompanying "trade winds" at lower latitudes. These flows are similar to their terrestrial weather phenomena of the same name, although on a much larger scale. The jet stream occurs at about 20,000 km below the surface and is 30,000 km across, while the lower latitude flows split into distinct bands north and south of the equator, 60,000 km across. The jet stream has plasma flows of order 10% (or 120 km/hr) faster than their surroundings while the trade wind flows are 16 km/hr faster.

There is also evidence for a strong polar flow (meridional flow) from equator to pole in each hemisphere. These polar flows are only about 80 km/hr, which is about eighty times smaller than the average solar rotation speed. At the poles, the flows turn vertical and flow into the solar interior, where they turn again and head towards the equator. At the equator the flows rise to the surface again and the process continues. These meridional flows help regulate the length of the solar cycle and explain the equatorward drift of the patterns seen in the butterfly diagram (Spörer's law).

High-resolution helioseismic observations of sunspots have also returned some surprises. The internal structure of sunspots can now be explored in great detail. Just under the sunspot (~3000 km) the internal flows slow down relative to the rest of the Sun, with the plasma flowing into the spot. Further beneath the sunspot (6,000–9,000 km) the flows are faster than average and there is a distinct outflow from the spot. These observations suggest that the deep sunspot region is hotter than the surrounding plasma, although this is inconclusive at the present time. In addition to the horizontal flows, the flow maps beneath sunspots also show a ring of strong downwards-flowing plasma around the sunspot near the top of the convection zone, with a similar ring of upwards flowing plasma at greater depths. The pattern of flows observed suggests that sunspots have a root-like structure rather than being a single monolithic tube of magnetic field—that is, just below the surface the sunspot separates into many individual magnetic tubes like the roots of a tree. The inward flows near the top of the spot hold the different roots together to form the spot.

An exciting application of local helioseismology is the ability to observe sunspots emerging on the far side of the Sun. The presence of a sunspot close to, or on, the surface on the opposite side of the Sun from the Earth causes the wave travel time to be shorter by as much as six seconds. This is important as it allows solar scientists to learn about conditions on the whole Sun rather than just on the side they can see. Such a capability has major implications for understanding the global nature of the Sun's magnetic field and for predicting space weather.

SOLAR NEUTRINO PROBLEM

One of the strangest by-products of the nuclear fusion process that make stars shine (see Chapter 2) is a tiny particle, called a neutrino or "little neutral one" by famous Austrian-Swiss physicist Wolfgang Pauli (1900–1958), that does not interact very well with ordinary matter. As a consequence, neutrinos, once produced, pass unhindered through the Sun, the Earth, and pretty much everything else. Stick up your thumb and roughly 5 billion solar neutrinos pass through it every second. Neutrinos come in three different families: the electron-neutrino, the mu-neutrino, and the tau-neutrino. The three families of neutrinos differ by the masses of the particles. These masses are difficult to measure, but particle physics experiments can place constraints on the mass. The masses of the neutrinos are so small that we shall compare them to the mass of the proton ($m_p = 1.67 \times 10^{-27}$ kg). Latest measurements suggest that the electron-neutrino mass is less than 2.3×10^{-9} m_p, the mu-neutrino mass is less than 1.8×10^{-4} m_p, and the tau-neutrino mass is less than 1.65×10^{-2} m_p.

Since neutrinos pass straight through the Sun after being released in a nuclear reaction, they can act as probes of the core of the Sun—their numbers and energies tell us something about the reaction rates, temperatures, and densities in the solar core. If these neutrinos could be detected on Earth, we could probe the nuclear furnace of the Sun.

As discussed in Chapter 2, the radiation of the Sun is produced primarily when protons collide in the hot, dense core, ultimately to produce helium. The sequence of nuclear reactions starting with two colliding protons and resulting in a helium nucleus is known as the p-p chain. Written as a sequence of equations, the p-p chain can be described as follows:

$p + p \rightarrow {}^2H + e^+ + \nu_e$ or *proton + proton → deuterium nucleus + positron + a neutrino*
$^2H + p \rightarrow {}^3He + \gamma$ or *deuterium nucleus + proton → helium-3 nucleus + energy*
$^3He + {}^3He \rightarrow {}^4He + 2p$ or *helium-3 + helium-3 → helium-4 nucleus + 2 protons*

where a positron is an anti-electron. It has the same mass as an electron, but the charge is positive rather than negative. The subscript "e" in ν_e indicates that the neutrino is an electron-neutrino. In the p-p chain, the neutrinos are produced in the first reaction of two colliding protons. Approximately 10^{36} of these reactions occur every second. The typical energy of the neutrinos produced in the p-p chain is 0–0.4 MeV. Another important neutrino producing reaction is that of beryllium being converted into boron:

$^7Be + p \rightarrow {}^8B + \gamma$ or *beryllium atom + proton → boron atom + energy*
$^8B \rightarrow {}^8Be + e^+ + \nu_e$ or *boron atom decays to beryllium atom + positron + neutrino*

The typical energy of the neutrinos produced in this reaction is 0–15 MeV. There are a number of other reactions that produce neutrinos at different rates, and energies, but the p-p chain is the most dominant and the Be → B sequence is the major contributor to the experimental detection discussed in the next section.

....

Energy Units

When dealing with fundamental particles like protons, electrons, and neutrinos, the energies are so small that our usual unit of Joules becomes cumbersome. To circumvent this, a new set of units was developed as a short cut. For small amounts of energy, scientists use the electron-Volt (eV), which is a measure of energy defined by the amount of energy equivalent to that gained by a single electron when it is accelerated through an electrostatic potential difference of one Volt, in a vacuum. One electron-Volt is equivalent to 1.6×10^{-19} Joules. As further shorthand, 1 kiloelectron-Volt (1 keV) denotes 1,000 electron-Volts and 1 megaelectron-Volt (1 MeV) denotes one million electron-Volts.

....

So what is the neutrino problem? The weakly interacting nature of neutrinos means that of all the billions of neutrinos to pass though every square centimeter of the Earth only a handful of them will take part in a reaction. In 1964, American scientists Ray Davis Jr. (1914–2006) and John Bahcall

(1934–2005) built the first solar neutrino detector; Ray Davis Jr. shared the 2002 Nobel Prize in Physics for his work. This detector consisted of over 600 tons of dry-cleaning fluid (perchloroethylene) located in the Homestake gold mine 1480 m below the mountains of South Dakota. The chlorine atoms in the perchloroethylene interact with neutrinos to produce the element argon:

$^{37}Cl + \nu_e \rightarrow ^{37}Ar + e^-$ or *one chlorine atom + a neutrino → one argon atom + an electron.*

This experiment can only detect electron-neutrinos and only if they have energy greater than 0.86 MeV, making it sensitive only to the neutrinos produced in the Be → B reaction. By measuring the amount and rate at which argon atoms are produced in the tank, one can gain a measurement of the number of neutrinos passing through the tank. Locating the detector deep underground minimizes contamination from interactions of cosmic rays with the Earth's atmosphere. Given the best models of nuclear fusion in the Sun's core and the probability that a passing neutrino will interact with a chlorine atom in a 615 ton tank (containing more than 10^{30} chlorine atoms), the Homestake detector was expected to detect between 235 and 236 neutrinos in a year. The earliest measurements taken by the Homestake mine experiment detected about one-third of this number. This smaller-than-expected detection is usually referred to as the *solar neutrino problem.*

Since the Homestake experiment, a number of other neutrino detectors have been developed, covering the wide range of neutrino energies produced in the solar fusion reactions. All of them measure lower-than-expected neutrino fluxes.

The solar neutrino problem means one of two things: either the models of the conditions in the solar interior are wrong, or the standard model of particle physics, which describes the properties of neutrinos, is wrong. The detailed observations provided by helioseismology, studying solar oscillations, have shown that our model of the Sun is extremely good. The data are so accurate that a very small change in the properties of the Sun's core would quickly result in a significant deviation from the observations. We are left with the conclusion that the fundamental description of particle physics did not accurately reflect the properties of the neutrino. Where once the neutrino was assumed to be massless, the solar observations have shown that they actually have a small mass that allows them to oscillate between the different families of neutrinos on their journey to the Earth—for example, an electron-neutrino to a tau-neutrino. Thus, any electron neutrinos that transformed into a mu- or tau-neutrino would escape detection by most of the neutrino detectors. Two recent experiments, the Super-Kamiokande in Japan and the Sudbury Neutrino Observatory in England, can detect the different neutrino families and have confirmed that the neutrinos do indeed oscillate between different kinds. The solar neutrino problem has been solved!

FUTURE OUTLOOK

Over the last two decades, the study of helioseismology has produced major breakthroughs in our understanding of the physics of the Sun, including the fundamental physics discovery that neutrinos can change between types. The next two decades look even more promising as new analysis techniques are developed, longer time series are produced, and new dedicated missions come into operation. These new and improved observations and techniques will significantly advance our understanding of what powers the solar cycle and how and where magnetic field is generated.

In August 2009, the National Aeronautics and Space Administration (NASA) will launch its first major Living with a Star space mission, called the Solar Dynamics Observatory (SDO). One of the instruments on SDO will be a high-resolution helioseismology telescope called the Helioseismic and Magnetic Imager (HMI), which will image the whole solar disk with sixteen times the resolution of the SOI/MDI telescope on SOHO. HMI will measure the motion of the solar surface to study solar oscillations in detail never seen before. One of the biggest breakthroughs expected is in the area of local helioseismology where the high resolution will enable HMI to probe the subsurface structures and flow down to much smaller scales.

In 2013, the European Space Agency (ESA) plans to launch the Solar Orbiter, which has the distinguishing characteristic that it will leave the ecliptic plane (the plane in which all of the planets, including the Earth, lie) to provide a view of the Sun from a high latitude. This new perspective will yield helioseismic observations near the poles of the Sun, a region not well covered from the Earth or from SOHO or SDO. When combined with data from SDO and ground-based observatories, the Solar Orbiter observations will complete the coverage of the whole Sun, allowing us to probe deeper into the solar convection zone to provide a 3D map of the motions and structure throughout the convection zone.

Looking inside the Sun is a challenging area of science, but the reward for the efforts of many solar scientists around the world has been great. The scientific discoveries made since that first tentative observation in 1962 have altered our perspectives on the physics of stars and, in fact, through the solution of the neutrino problem, to fundamental physics itself. Like a good whisky or an original first-edition comic book, helioseismology only gets better with age. The longer we observe, the more modes we can pull out of the signal, the more information we gather about the subsurface motions. As new instrumentation, improved computer models, and better analysis techniques get developed, the prospects only look brighter and brighter.

Oscillations on Other Stars

Observing oscillations on the Sun has proven to be a powerful tool for understanding the inner workings of our nearest star. The relative proximity of the Sun allows us to see these oscillations in great detail and to develop techniques that extract the relevant information regarding the structure and flow patterns beneath the Sun's surface. The Sun should not be unique in this respect, and one would expect that other stars should show signs of oscillatory behavior. The sheer distance to the stars means that the information on such oscillations would be extremely difficult to obtain. However, some of the global mode solar techniques are now being applied to stars and a new field of science called *asteroseismology* is developing.

Many different kinds of stars show large-scale coherent oscillations of the type exhibited by the Sun. Many stars have been observed to show pulsations where their brightness varies on a regular pattern. These pulsations act like the global "breathing" modes of the Sun. While we can observe the Sun in great detail and infer a lot about stars in general from these observations, the ability to detect multiple modes of oscillation on other stars will provide a great leap forward in our understanding of how stars work and the relationship between their internal structure and their activity. This, in turn, will teach us more about the Sun itself. Asteroseismology provides information about the mass and age of the star with great precision revolutionizing our understanding of how stars evolve. The science of asteroseismology is expected to develop rapidly over the next five to ten years as ground-based observational studies improve and new space-based missions gather sufficient data to study stellar oscillations. For example, results are coming in from the Canadian MOST (Microvariability and Oscillations of STars) mission, which was launched in June 2003. Among several interesting scientific results, MOST has found stars that exhibit multiple oscillation frequencies and has studied starspots to determine the rotation patterns of nearby stars and found a similar behavior to that of the Sun. The French mission, CoRoT (Convection, Rotation, and planetary Transits), launched in December 2006, is expected to produce its first detailed observations of stellar oscillations within the next few months.

Finally, NASA's Kepler mission, to be launched in spring 2009, is designed to search for Earth-sized planets around other stars. As part of this study, Kepler will study stellar oscillations to determine the ages and masses of the stars. All of these missions and ground-based programs promise to advance our understanding of asteroseismology.

RECOMMENDED READING

Zirker, John B. *Sunquakes: Probing the Interior of the Sun.* Baltimore: Johns Hopkins University Press, 2003.

WEB SITES

asteroseismology: http://asteroseismology.org.
GONG: http://gong.nso.edu.
Helioseismology: http://solar-center.stanford.edu/about/helioseismology.html.
SDO/HMI instrument page: http://hmi.stanford.edu.
SOHO/MDI instrument page: http://soi.stanford.edu.
Wave Motion Basics: http://www.school-for-champions.com/science/waves.htm.

5

Solar Acne: Sunspots and the Solar Surface

Our standard view of the Sun is of a shining circular disk with a well-defined sharp edge. Because the Sun is a big ball of gas, there is no solid surface like there is on Earth. However, the rapidly changing density of the gas means that light from just below the surface gets absorbed by the gas before it can reach the surface and "escape" into space. The sharp edge marks what solar scientists call the solar surface, or more technically, the photosphere, from the Greek word *photos* meaning light. When one looks closely at the surface of the Sun, a thing you should never do without special filtering equipment, it is far from smooth and uniform. The Sun has a dark rim around its edges, the surface is occasionally pock-marked with dark "imperfections" or sunspots, and the surface itself seems to be tiled with irregularly shaped features that roil and roll. Without these imperfections and features we would know very little about the Sun and may never have come to shake the belief that the Sun orbited the Earth.

This chapter describes the photosphere and its properties, with a major focus on sunspots and what we have learned from them. Sunspots, in particular, are extremely important in understanding the Sun and its behavior. They yield insight into the inner workings of the Sun through their formation, evolution, and regular patterns of change (Chapter 6), while their strong magnetic fields provide much of the energy for heating the million-degree solar atmosphere (Chapter 7) and for explosive solar storms that can affect the Earth (Chapter 9).

Observing the Sun

One should never look directly at the Sun, not even during an eclipse or at sunrise or sunset. The Sun emits high-energy ultraviolet (UV) radiation that can permanently damage your eyes and, in extreme cases, even cause total blindness. The safest way to observe the Sun is by using specially developed Mylar sunglasses, through a telescope fitted with a solar filter that blocks most of the sunlight, or best of all, by projecting an image of the Sun onto a piece of white card. The simplest way to do this at home is to build a pinhole projector. Here are a few easy steps to start observing the Sun:

Take a piece of stiff card, and cut a hole about the size of a quarter out of the center. Tape some flat (i.e., no creases) aluminum foil over the hole, and poke the smallest hole you can in the center of the foil with a fine pin or needle. Hold the board so the Sun's light can pass through the pinhole you created in the foil. DO NOT LOOK AT THE SUN THROUGH THE HOLE! Hold a piece of white card or paper behind the pinhole card to see a small bright dot. This bright dot is an image of the Sun. The further you separate the white board from the pinhole board, the larger the solar image will be (you might need a friend to help with this). If you are lucky you might see some sunspots.

It is frequently, but incorrectly, stated that sunspots were discovered by the famous Italian scientist Galileo Galilei (1564–1642) who pioneered the use of the telescope, a then new instrument invented in Holland in 1608, for astronomy. Naked eye observations of spots on the Sun have been reported independently by Chinese and Greek astronomers from as far back as the fourth century BC, with sporadic mentions throughout the subsequent centuries by Arabic, Chinese, and European astronomers. Of particular note is a record that appears in the first edition of the great Chinese Encyclopedia, published in 1322, which described observations of forty-five sunspots between the years AD 301 and 1275. The study of sunspots became systematic when telescopic observations became common, with several astronomers claiming to have seen sunspots in the year 1611. Galileo claimed to have been the first to discover sunspots, but published works had appeared prior to his first "official" recording of his observations. Astronomers Johan Goldsmid (1587–1616; Holland), Christopher Scheiner (1575–1650; Germany), and Thomas Harriot (1560–1621; England), along with Galileo, all share the honor of the first telescopic observations of sunspots. While there is considerable debate as to who saw the sunspots first, it was Goldsmid (also known more famously by his Latin name Fabricius) who first published reports of his observations in his book *De maculis in sole* (1611). Despite his detailed observations, there was no consensus on what the sunspots actually were, with Galileo and Fabricius believing them to be cloud-like phenomena on the Sun itself; Scheiner, who was also a Jesuit priest, thought them to be moons or planets orbiting the Sun. Galileo argued his case based on a series of sunspot observations he made over several weeks. He saw that as the sunspots approached the edge of the Sun they would appear foreshortened, which could only happen if the sunspots lay flat on the solar surface.

The reason that this was a critical observation with major significance for future developments in astronomy was that it raised questions about the doctrines that had been set in place by Aristotle in the fourth century BC and adopted by the Catholic Church, which dominated scientific thought in Europe throughout the Middle Ages. This doctrine specified that all objects in the heavens were divine in nature and therefore perfect. Sunspots represented blemishes on the face of the Sun and thus demonstrated that heavenly bodies were not as Aristotle had claimed. These telescopic observations, along with the discovery of craters and mountains on the surface of the Moon, the moons of Jupiter, and the phases of Venus, led to a wider acceptance of the heliocentric universe model of Copernicus (Chapter 1) and set us on the path to modern-day astronomy.

In the intervening four hundred years, since these early studies of sunspots, as telescopes improved and new instrumentation was developed, many discoveries were made that ultimately led to our modern-day understanding of the Sun. Discoveries as basic as the facts that the Sun rotates, that the Sun is a star, and that it has a magnetic field all came about from observing the photosphere and its sunspots. The discovery of granulation, the solar cycle, and the latitudinal variation of the solar rotation all led to major advances in understanding how stars work. In this chapter, we will describe where these discoveries have taken us in our quest to understand the Sun.

THE SOLAR PHOTOSPHERE

The solar photosphere is the deepest layer that we can see directly. The photosphere is a little over 400 km thick and marks the transition from where the solar plasma is almost completely transparent to where it is almost completely opaque. Light generated below the photosphere gets readily absorbed by the higher layers and does not make it out into space. The photosphere is, therefore, the visible face of the Sun, the face that we see when we project sunlight through a pinhole (see earlier sidebar). Prior to the invention of spectrographs (see Chapter 2) that show the different temperature layers of the Sun, coronagraphs (see Chapter 7) that block out the main solar disk to show the faint solar corona, and telescopes to view the Sun across the electromagnetic spectrum (radio, X-rays, UV, etc.), the photosphere was the only region we could observe directly; much of our earlier information about the Sun came from a study of this very narrow layer (400 km corresponds to less than six hundredths of a percent of the solar radius).

The rapid transition from transparent to opaque in the solar photosphere is a result of the fact that as the temperature increases with depth it traverses the temperature at which hydrogen, the most abundant material in the Sun, is completely ionized, that is, the temperature is sufficiently high that the electrons in the hydrogen atoms get kicked off the atom via "collisions" with the radiation associated with that temperature. This is crucial since free electrons are extremely good at scattering radiation and preventing it from penetrating through the plasma. As one goes higher in the photosphere, the temperature cools such that the radiation no longer has enough energy to ionize the hydrogen and the electrons remain confined to the atoms. With no electrons to interact with, the photons travel unimpeded through the remaining layers of the photosphere and out into space. The sharpness of this transition from a predominantly ionized hydrogen layer to a predominantly non-ionized (neutral) layer gives the Sun its well-defined edge. This sharp change from opaque to transparent delineates a marked transition from the solar interior to the solar atmosphere, and, consequently, solar scientists typically refer to this layer of the photosphere as the surface of the Sun.

Study of the photosphere led to a basic understanding of many of the large-scale processes that drive much of the solar behavior. Because the transition from transparent to opaque depends upon the wavelength of the light being considered, much can be determined about the physical conditions of the photosphere, such as temperature, density, and chemical composition. Detailed studies of the photospheric spectral lines (see Chapter 2) and how their intensity varies across the solar disk led to the discovery of many dynamic solar phenomena, one of the most important of those being granulation, which we will discuss later in this chapter.

The general properties of the solar photosphere are summarized in Table 5.1, where we see that the temperature decreases steadily with height from the deepest layers at ~100 km to the temperature minimum region (T = 4400 K) which lies at a height of approximately 500 km. On this scale, the visible edge of the Sun, the solar limb, lies at a height of 350 km. Above the temperature minimum region, the temperatures rise dramatically to the values measured in the chromosphere and corona (see Chapter 7). The total mass and hydrogen number density increase by a factor of ~50

Table 5.1. **Properties of the solar photosphere as a function of depth**

Height (km)	Tempera- ture (K)	Mass density (kg/m^3)	Hydrogen number density (m^{-3})	Electron number density (m^{-3})
503	4400	6.22×10^{-6}	2.65×10^{21}	3.04×10^{17}
378	4610	1.92×10^{-5}	8.19×10^{21}	9.13×10^{17}
200	4990	8.20×10^{-5}	3.50×10^{22}	4.00×10^{18}
100	5410	1.67×10^{-4}	7.14×10^{22}	1.00×10^{19}
0	6520	2.78×10^{-4}	1.19×10^{23}	7.68×10^{19}
−100	9400	3.18×10^{-4}	1.36×10^{23}	3.86×10^{21}

Source: Adapted from the model of Maltby et al. (1986)

over the 600 km scale of the photosphere, while the electron number density increases steadily by a factor of ~250 over the first 500 km before jumping by an additional factor of 50 in the last 100 km.

....

The Solar Photosphere and the Big Bang

How could we possibly relate the surface of a star to the big bang itself? Well, remarkably the physics that makes the solar surface so sharply defined is exactly the same as that which defines the last scattering surface associated with the production of the cosmic microwave background and the theory of the Big Bang. As the universe expanded from the initial explosion of the Big Bang, it rapidly cooled. The major constituents of this early universe were hydrogen and helium (with a little lithium thrown in). At the higher temperatures of the early Big Bang, hydrogen was completely ionized and the electrons associated with the hydrogen atoms were unconfined. The radiation associated with these high temperatures was continually scattered by these free electrons and could not travel far without encountering a scattering electron. As the continuous expansion cooled the early universe, the electrons recombined with their parent hydrogen nuclei to form a predominantly neutral hydrogen universe. Like in the solar photosphere the radiation could now travel unhindered throughout the whole universe, generating a cool background that we now know as the cosmic microwave background that lies at the hub of the Big Bang explanation for the universe.

....

Observations of the solar photosphere rely on the total amount of visible light emanating from the surface—these are usually referred to as white light observations as they integrate the emission across the visible spectrum. The upper reaches of the photosphere near the temperature minimum region (~400–500 km above the white light height) also emit UV radiation, and so telescopes sensitive to wavelengths centered around 160 nm are used. Even over such a short range of heights the structure of the photosphere changes dramatically. The granulation patterns so evident in the white light data are replaced with a network of bright localized structures and the sunspot regions are surrounded by sources of bright emission called faculae. This dramatic change in the observed structure is a result of how the energy is transported out into the solar atmosphere and the influence of magnetic fields. When observing the solar photosphere, we notice two distinctly different regimes: in and around sunspots strong magnetic fields dominate, and we call this the *active Sun*; everywhere else behaves more or less in the same fashion and is known as the *quiet Sun*. Note that even in the quiet Sun regions, the photosphere is constantly changing; nowhere is the Sun actually "quiet." We will discuss sunspots and their strong magnetic fields in the next section. For now, we will concentrate on the quiet Sun and discuss the large-scale patterns of motion known as granulation.

Close inspection of the solar surface shows a highly organized structure with the surface broken up into bright cells of varying sizes, separated by dark lanes. This pattern is known as granulation and provides a direct and crucial clue to the dynamics of the solar interior just below the surface (what we now know as the convection zone; see Chapter 2). The

Limb Darkening

The picture of the solar photosphere shows a clear reduction in the intensity towards the solar limb. This is a consequence of the spherical nature of the Sun and the structure of the photosphere itself. The photosphere is a very thin layer that penetrates a few hundred kilometers into the Sun. As we go deeper into the photosphere, the temperature increases, resulting in a corresponding increase in the intensity of the light produced. When we look at the center of the solar disk, the light we see at the Earth is dominated by the deeper, brighter layers of the photosphere as the light takes the shortest path to our eye. Near the edge of the Sun, the solar limb, the light from the lower depths has to travel at a shallower angle to reach the Earth and so must pass through more of the

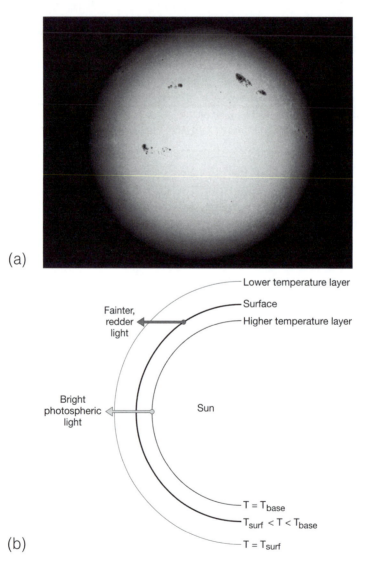

Figure 5.1 Limb darkening: (a) Optical image of Sun showing the darkening of the light towards the edges of the solar disk and (b) schematic showing reason for limb darkening. Illustrations by Jeff Dixon.

upper layers on its way out of the photosphere. This results in more of the radiation from the deeper, hotter regions being absorbed, so the light we see is dominated by the higher, cooler, and thus less intense upper regions of the photosphere. This reduction in intensity results in the limb-darkening effect seen in the images. Another artifact of the dominant layer of the photosphere changing as we go from disk center to the limb is that the limb also appears to be redder, associated with the lower temperatures.

....

characteristic size of these bright granules is about the size of Texas ($\sim$1000 km across) with their centers being separated by about 1,400 km (the dark lanes separating them are very narrow $\sim$200–400 km). What is even more remarkable is that this mosaic is continually varying in time. In the center of the bright granules the plasma moves upwards, while in the surrounding dark lanes the plasma flows downwards. Because most of the photospheric observations are taken from the ground, the absolute values of the velocities are difficult to determine against the image motions generated by the wiggling of the Earth's atmosphere. In addition, the spectral lines used to measure these motions are complicated and vary across each granule, and these motions are superimposed on top of the five-minute oscillations (see Chapter 4), again making an accurate velocity determination difficult. However, one can measure the relative velocity difference between the upflows in the centers and the downflows at the edges of the granules. This is found to be of the order of 1.8 km/s ($\sim$6,500 km/hr)—in other words, the plasma flows back down into the Sun with a speed of almost 2 km/s slower than it flows out. In addition to these vertical flows, the granulation cells show a horizontal flow of plasma from the center to the edges of approximately 1.4 km/s. These flows tell us a lot about the physics of granulation and can be related through advanced computer modeling to both the temperature structure and overall evolution of the cells. Such modeling has shown that the actual temperature difference between the hot upflowing plasma and the cool downflowing plasma can be as much as 5000 K, and that the downflows are generated by rapid cooling due to the loss of energy as radiation.

The granular cells are thought to form from several smaller structures, expand to their maximum size, and then split up again into several pieces and fade from the observations as they do so. The actual granules themselves have an average lifetime of only 5–10 minutes. The picture that emerges is of a highly dynamic photosphere with material flowing outwards from below, "spilling" horizontally over a scale of about 1,000 km, and then flowing back into the interior. This is very similar to the pattern one would see if making porridge or soup where the pan is heated from below, the hot soup rises and cools, and falls back down to be heated again. This convection gradually spreads the heat from the stove to the contents of the pot. Looking down from above, the soup would appear to be broken up into individual cells of upwelling material just like the surface of the Sun (see sidebar on convective motions).

Convective Motions

Figure 5.2 shows the implications of the motions and temperatures observed in surface granulation on the Sun. This is the standard picture of convection motions assumed to be responsible for creating weather patterns in the Earth's atmosphere (e.g., trade winds), tectonic motions on the Earth's surface (mantle convection), the pattern of brown-and-white bands on the planet Jupiter, and many other similar phenomena. It merely reflects the behavior of a fluid that is heated from below. This process is called convection and is used every day in kitchens and campfires around the world.

The layer of fluid at the bottom, nearest the source of energy, gets heated first. The increase in temperature increases the pressure in that layer and the fluid, causing it to expand. The expansion in turn reduces the density of the fluid in that layer, making it buoyant, and so it rises. The vacancy left by the rising fluid is filled by denser plasma from above, which, in turn, gets heated and rises. The rising fluid shares its excess energy with its surroundings and so cools, contracts, becomes more dense, and then falls, to repeat the cycle again when it reaches the heat source. The rising and cooling breaks the fluid up into cells, which essentially act like a conveyor belt, replacing the cooler material above with newly heated material from below.

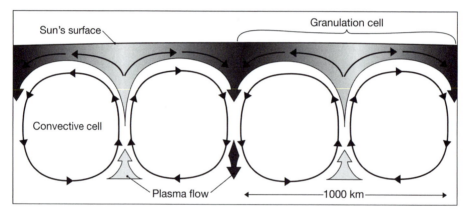

Figure 5.2 Mechanism of solar convection and granulation. Upflowing and downflowing plasma occurs between the convection cells, while plasma flows horizontally outwards from center of cells. Illustration by Jeff Dixon.

The granulation pattern is seen across the whole solar surface, but changes significantly in the vicinity of the strong magnetic fields of sunspots. Near sunspots, the granulation cells are smaller by about 20% (i.e., a typical size of ~800 km) and their brightness contrast is about half what it is in the quiet Sun.

A picture of the solar disk showing Doppler velocities, where the upflowing and downflowing plasma appear as brightness variations, shows a large-scale pattern on a scale some thirty times larger than that of the granulation motions. This supergranulation is made of giant cells of the order 30,000 km across (more than twice the size of the Earth) with plasma flowing across the surface at about 0.5 km/s (or 1,800 km/hr). Like the granulation cells, there is evidence for upflows in the center and downflows at the edges of these cells. The lifetime of these cells is not known exactly, but they appear to last around twenty-four hours.

The quiet Sun is a sea of undulating motions with the large convection cells appearing as a granular pattern on the solar surface. These motions and their coupling to the Sun's magnetic field provide much of the energy for heating the quiet solar atmosphere and driving the solar wind. These topics will be discussed in later chapters. While this activity is impressive, the solar photosphere is home to the granddaddy of solar activity, the sunspot, where intense magnetic fields provide the engine for some of the most energetic phenomena in the solar system.

SUNSPOTS

At the beginning of the chapter, we discussed briefly the historical context of solar observations, and sunspots played a key role. Observations of sunspots since 1611 have been fundamental in understanding solar activity, solar magnetism, solar variability, and the impact of the Sun on Earth's climate. The development, structure, location, and number of sunspots all provide crucial information on the inner workings of the Sun and the connections between happenings in the interior, and their subsequent effects in interplanetary space and their interaction with the Earth. In the next few sections, we will discuss the observational underpinnings of sunspots and what they have allowed us to understand about the physical behavior of the Sun.

··

Wilson Depression

An interesting feature of sunspots is that their shape is observed to flatten significantly as they rotate towards the solar limb. This effect was first noticed by Scottish astronomer Alexander Wilson (1714–1786) in 1769. Wilson's observations also showed that the penumbra of the spots on the side nearer the center of the disk appeared narrower, while the penumbra on the other side nearer the solar limb remained at roughly constant width. Wilson pointed out that behavior might be explained if the spots were actually plate-shaped depressions in the photosphere. The Wilson depression has been estimated to be as much as 600 km, but the observations are not regarded as being reliable.

··

Sunspots are instantly recognizable as dark blemishes on the bright surface of the Sun, but a closer look shows a surprising amount of structure and detail (see Figure 5.3). Sunspots first appear on the solar disk as single, small dark regions, called pores, which grow larger and darker while developing a surrounding annulus of alternating bright and dark filaments. The dark central area is called the *umbra* (Latin for shadow) and the outer ring is called the *penumbra* (almost-shadow). Typically, the umbral portion of a sunspot has a diameter of around 10,000 km, with the penumbra extending a further 5,000 km or so. The idealized sunspot is a dark circular umbra surrounded by a brighter circular ring-like penumbra. Unfortunately, very few sunspots are ideal. Large sunspots may develop without penumbrae, sunspots appear in complex groups with oddly shaped umbrae and shared penumbrae, and penumbrae can even appear in the absence of any umbrae. To help distinguish the different kinds of sunspot, solar scientists have developed a number of classification systems, including the Waldmeier, the McIntosh, and the Mt. Wilson classifications. The Waldmeier (or Zurich) classification scheme was introduced in 1938 by Swiss solar physicist Max Waldmeier (1912–2000) and defines a sunspot by the sign of its magnetic field, how many individual spots make up the group, the characteristics of the penumbra, and its size. While used heavily by amateur astronomers, this scheme does not allow for the range of complexity exhibited by sunspots. In 1990, U.S. astronomer Pat McIntosh introduced a modified scheme in which sunspot complexity is categorized by a letter denoting the form of the penumbra. One of the most commonly used classification schemes was developed by the Mount Wilson Observatory in Pasadena, California, which uses measurements of the strength and distribution of the sunspot magnetic field. This scheme was first introduced by George Ellery Hale and Seth Barnes Nicholson in 1938.

The magnetic complexity of a sunspot is a useful criterion for understanding the level of activity, the production of solar flares, and the heating of the surrounding atmosphere. The Mount Wilson classification defines magnetic configuration in three ways: *Unipolar* (denoted α) where only one magnetic polarity is present, *Bipolar* (denoted β) where the spots appear as

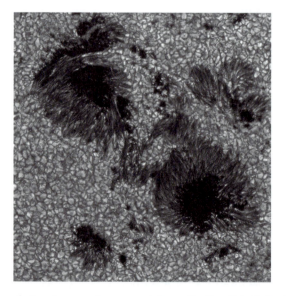

Figure 5.3 High-resolution image of a sunspot from the Swedish Solar Telescope on La Palma. Courtesy of Göran Scharmer and Kai Langhans, Institute for Solar Physics of the Royal Swedish Academy of Sciences, Sweden.

a pair of opposite polarity (although a bipolar region could include multiple spots), and *Complex* (denoted γ) where the sunspot groups are irregularly distributed. The full Mount Wilson classification is as follows:

UNIPOLAR GROUPS

A = α	symmetrical distribution between leading and following part of group	
AP = αp	leading part of group is elongated	
AF = αf	following part of group is elongated	

BIPOLAR GROUPS

B = β	leading and following polarities have roughly equal area
BP = βp	preceding polarity is larger component of group
BF = βf	following polarity is larger component of group

COMPLEX BIPOLAR GROUPS

BG = βγ	generally bipolar but no clear north-south divide between opposite polarities
G = γ	complex distributions of mixed polarities
D = δ	spots of opposite polarity are close together and share a penumbra

The Mount Wilson magnetic classification and the McIntosh optical classification serve a major purpose in understanding how different sunspot configurations relate to different levels of activity and are used to determine space weather alerts and warnings (Chapter 10).

....

Why Are Sunspots Dark?

One might think that sunspots are dark because they are cold and don't emit much light. However, if one could physically remove a spot from the Sun (maintaining all of its attributes), the sunspot would shine brightly enough to blind you even at 150 million kilometers distance. Sunspots have temperatures of around 4200 K, which is about 1600 K cooler than the surrounding photosphere. The strong dependence of emission on temperature (brightness is proportional to the fourth power of the temperature) means that sunspots are relatively faint (about 25% as bright) compared to the photosphere, and in contrast show up as dark against the blinding brightness of the rest of the Sun.

The real question then is not "why are sunspots dark?" but why are they so cool? The basic idea is that the presence of the strong magnetic field extending deep into the convection zone prevents the normal process of convection from working in sunspots (see earlier sidebar). The suppression of the convection limits the amount of energy that can be transported to the surface, yielding a lower temperature for the sunspot.

....

There are several theories for how sunspots form: one states that superganulation motions push magnetic field together to form an intense concentration; another model, the *cluster model*, assumes that the sunspot is a loose collection of small magnetic field concentrations that extend below the surface like the roots of a tree (the sunspot being the trunk). However, recent developments in helioseismology (Chapter 4) suggest that magnetic field forms at the base of the convection zone as a large elongated concentration, known as a *flux tube*. The strong field of this flux tube increases its pressure relative to the background, making it buoyant like a hot air balloon and causing it to rise to the surface where it pops through the surface as a sunspot.

During their lifetime, which may be hours, days or even months, sunspots exhibit a variety of behavior. The development of a sunspot or sunspot group, how fast it grows and decays, and how it varies during its lifetime provide important clues to the origin of sunspots in the solar interior. Sunspots generally appear in groups of two or more; single spot groups are very rare. Sunspots first appear as a pore, a region of enhanced magnetic field, on the surface of the Sun, which continues to grow over a period of a few days. A sunspot group is a collection of one or more pores that develop together. During its early growth, the sunspot group spreads out, generally in an east-west direction, to reach some maximum area. During this expansion, some of the sunspots may coalesce to form larger spots, generating a penumbra in the process. The magnetic field continues to intensify and there is an associated increase of the faculae, the name given to the regions of bright hydrogen emission around sunspots. After the first few days, the area of the sunspot group maximizes and the region starts to decay, a process that may take days to months. The growth and decay of a sunspot depends on the individual sunspot, with some of the smaller ones both growing and decaying slowly, while the large long-lived ones tend to grow rapidly but decay slowly. Sunspot decay is characterized by the outward drift of small portions of the umbral magnetic field. This reduces the sunspot area and gradually the sunspot "erodes" away. The magnetic field leaving the sunspot is assumed to cancel with opposite polarity field in the surrounding photosphere.

The whole process of growth and decay of a sunspot or sunspot group can take anywhere from hours to months, with more than 50% of sunspots lasting two days or less. Only 10% of sunspots survive more than eleven days, and very few last more than a couple of months. Roughly speaking, the larger the area of a sunspot the longer it will live with the approximate relation: $T = A_S/10$, where T is the lifetime in days and A_S is the maximum sunspot area in millionths of the solar disk.

SUNSPOT MOTIONS

The life cycle of a sunspot is extremely interesting and provides crucial insight into their formation process and their impact on the solar atmosphere above them. The most obvious motion is that they apparently move across the solar disk (Figure 5.4), with their rate of motion varying with latitude on the Sun. These motions were evident in the earliest solar observations and demonstrated for the first time that the Sun itself rotated, and that it did so differentially (see Chapter 3).

As the sunspot grows, the magnetic complexity develops and this is reflected in the range of internal motions and plasma flows observed. The most evident plasma flow in sunspots occurs in the penumbra and was first detected by British scientist John Evershed (1864–1956) in 1909. When sunspots are observed near the solar limb, Evershed found that plasma from both the umbra and penumbra were moving radially outwards from the center of the spot and parallel to the Sun's surface. The flows were observed to accelerate outwards with a velocity of 2 km/s at the outer edges. Evershed also noted that hotter plasma, from the solar chromosphere, exhibited a tendency to flow inward. These flow patterns were subsequently confirmed and have been routinely measured over the intervening century. The pattern of sunspot flows is now known as the Evershed Effect.

The explanation for the Evershed effect and the differences between plasma at different temperatures is still not known. The original observations suggested that at lower heights the flow was outward, but as one looked higher in the photosphere and into the chromosphere, the flows slowed down and eventually reversed. Recent observations, however, have shown that the picture is much more complex, which at present defies a definitive explanation.

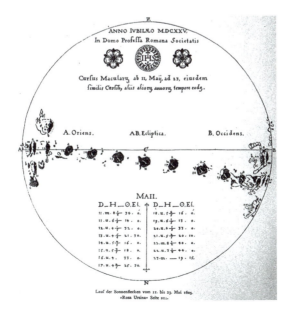

Figure 5.4 Passage of sunspots across the solar disk indicating that the Sun rotates, from Scheiner's *Rosa Ursina*. Courtesy of Rice University Galileo Project.

In addition to the radial flows in the penumbra, sunspots frequently exhibit a large-scale rotational motion about their umbral center. Sunspot rotation has been observed for more than a hundred years, but it has become more important recently as solar scientists try to understand what drives solar activity, particularly the large-scale eruptions known as flares and coronal mass ejections (Chapter 9). Sunspots have been found to rotate at a rate as high as three degrees per hour at the umbra-penumbra boundary about 7,000 km from the center of the spot. This translates to a velocity of about 100 m/s. Such rapid rotation has major consequences for solar activity as it represents the injection of a significant amount of energy into the magnetic field of the surrounding atmosphere. If one imagines the sunspot magnetic field as a large elastic band, the rotating spot then "winds up" this elastic band, adding energy. At some point the elastic cannot support the added energy and the elastic band snaps. In this way, it is believed that the dynamics of the sunspot wind up the magnetic field until the field "snaps" to release the energy as a solar flare or coronal mass ejection.

Finally, a point to note is that as helioseismic techniques improve (Chapter 4), plasma motions under sunspots are being detected with inward and outward flow patterns being found just below the surface, where the flows are inward, towards the center of the sunspot, for the first 3,000 km below the surface and outward below this. Local helioseismology observations are also finding signs of sunspot rotation below the surface, with rotational flows detected down to a depth of 5,000 km.

Sunspots are highly dynamic and because of their strong magnetic field play the principal role in driving much of the solar variability. Their behavior provides important diagnostics of the interaction between the solar interior, its surface, its atmosphere, and beyond to the Earth. A major clue to this interaction is provided by the array of recurring patterns that help define the role played by magnetic fields and the discovery of the solar cycle.

SUNSPOTS AND MAGNETIC FIELD

The magnetic field of the Sun was discussed in detail in Chapter 3, and our knowledge of solar magnetism stems from the observation of sunspots. The magnetic field is the defining property of a sunspot. The strong sunspot magnetic fields set them apart from the rest of the solar photosphere. Typically, sunspots consist of groupings of positive and negative polarity, which may take the form of a simple bipole with a concentrated and reasonably equal pairing of each polarity, or may be complex with multiple groupings of the two polarities in a mixed-field region. The magnetic field is most easily measured in the photosphere where it is the strongest. Sunspot magnetic fields have maximum values in the umbra of (2,000–3,500 G) dropping radially from sunspot center with penumbral fields in the range 700–1,000 G.

The generally bipolar nature of sunspots and the variation of the magnetic field suggest that the sunspots mark the intersection of a tube-shaped magnetic structure with the solar surface. This magnetic structure is called a *magnetic flux tube*. The basic idea is that an intense magnetic flux tube is generated by dynamo action at the base of the convection zone. The strong field makes it buoyant, and its central portion rises to the surface while the ends remain rooted where they formed. Because of the buoyancy the flux tube takes on an omega (Ω) shape and the top eventually pops out of the surface as a sunspot. As the flux tube continues to rise, the sunspot develops a bipolar structure with the two opposite polarities ("in to" and "out of" the surface) moving away from each other (the observed spreading of the sunspot region). The magnetic flux tube now has an interior part (observed by helioseismology), a surface part (the sunspots), and an atmospheric part where the flux tube rises above the surface to form a hot active region corona (see Chapter 7). It is the energy in the intense field of the sunspot that is responsible for many of the solar dynamic phenomena we will feature in later chapters.

SUNSPOT PATTERNS

As discussed in the preceding section, sunspots tend to come in groups with both positive and negative magnetic polarities. Relative to the direction of solar rotation the polarities are designated *preceding* or *following*, with the preceding polarity leading as they rotate with the Sun. While sunspot orientation is generally east-west, there are a number of interesting properties that serve to define a global pattern of development. These patterns and their relationship to the large-scale magnetic field of the Sun led to our current understanding of how and why the Sun varies, how and where it generates magnetic field, and what the effects of this variability are on the Earth. The observed sunspot patterns fit into five major categories that are listed here in rough chronological order of their discovery.

1. The Solar Cycle

The solar cycle is such an important observational phenomenon to all aspects of the Sun's behavior that it is the subject of the entire next chapter. For now, we briefly summarize the solar cycle within the context of the behavior of sunspots. One of the earliest discoveries in the investigation of sunspots was made by German pharmacist and amateur astronomer Heinrich Schwabe (1789–1875) in 1843. Having recorded the occurrences of sunspots for forty-three years, Schwabe noticed an unusual, but distinct, pattern: the number of sunspots varied with an approximately eleven-year periodicity; sunspot numbers would rise and decline on a regular eleven-year pattern (Figure 5.5). This pattern is known as the *solar cycle*.

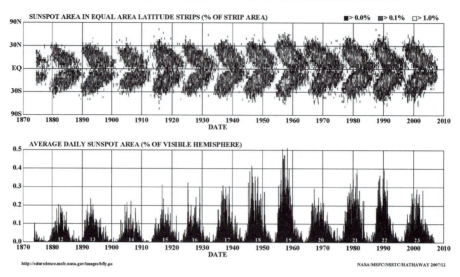

Figure 5.5 The butterfly diagram, showing the latitudinal distribution of sunspots over the course of several solar cycles. Courtesy of David Hathaway of NASA Marshall Space Flight Center.

Sunspot Number and Sunspot Area

A measure of solar activity is what is called the Wolf Number (or simply, the Sunspot number), introduced by Swiss astronomer Rudolf Wolf (1816–1893) in 1848. This was used as a means to test the sunspot cycle discovered in 1843 by German scientist Heinrich Schwabe. The Wolf number is characterized by the empirically derived relation:

$$R = k(10g + f)$$

where g is the number of sunspot groups on the visible face of the Sun, f is the total number of all spots in these groups, and k is a "reduction factor" to convert counts from different observations from different observatories to a single scale. The reduction factor, k, depends on a range of factors including atmospheric conditions at the observatory site, the characteristics of the telescope being used, and the level of solar activity. For example, if there are three sunspot groups on the Sun and each one has four sunspots, then R = 42 for k = 1.

Because solar activity is driven by the energy residing in the Sun's magnetic field, the number of sunspots alone is not a sufficient measure of the activity. To take account of this, the sunspot area number was devised to better represent the sunspot energy content on the solar disk.

The sunspot area number uses white light data from around the world to determine the sunspot areas for all sunspots visible on the Sun and combines them as daily averages. The record, compiled originally by the Royal Greenwich Observatory in England and now by the Mount Wilson Observatory, extends from the present day back to 1874. The sunspot area number, A, in units of millionths of the solar hemisphere is given by:

$$A = \frac{\left(\sum {A_s}/{\cos\theta}\right) \times 10^6}{2\pi r_0^2}$$

where A_S is the area of a given sunspot, the $\cos\theta$ term allows for projection effects, where θ is the solar longitude of the sunspot, r_0 is the measured radius of the Sun, and the 10^6 multiplier converts the fractional sunspot area to millionths of the solar disk.

The maximum magnetic flux density, B, in a sunspot (how much magnetic flux passes through a uniform area on the surface) is found to be related to the area of the spot, A_S, where B is measured in the magnetic units of Gauss and A_S is measured in millionths of the solar disk. The relationship can be written:

$$B = \frac{3700 A_s}{A_S + 60}$$

Thus, if sunspots cover 10 millionths of the solar disk, the maximum flux density is B = 528.6 G.

While this was a reasonably successful measure of solar activity, present day solar physicists use an array of other measures such as sunspot area, solar radio emission, and atmospheric UV and X-ray emission.

2. The Butterfly Diagram (Spörer's Law)

Excited by Schwabe's discovery, Englishman Richard Carrington (1826–1875), the son of a wealthy brewer, turned his attention to astronomy. (Richard Carrington was to become famous for another solar discovery, namely the first recorded observation of a solar flare; see Chapter 9). He meticulously observed sunspots from 1853 to 1861 and discovered another important pattern in their behavior. He found that sunspots appeared at different latitudes on the Sun and that, on average, the latitude varied with the solar cycle. Sunspots that appeared at the beginning of the solar cycle emerged at high latitudes, while end-of-cycle sunspots appeared closer to the solar equator, with a smooth transition in-between.

This latitude variation of sunspots was investigated in more detail by German astronomer Gustav Spörer (1822–1895), and Carrington's discovery is now known as Spörer's Law. A graphical representation of Spörer's Law is shown here in Figure 5.5, and for obvious reasons is known as the butterfly diagram. The butterfly diagram shows the eleven-year cycle, but also demonstrates that successive cycles overlap by one to two years.

3. Hale-Nicholson Polarity Law

As discussed in Chapter 3, George Ellery Hale discovered the presence of strong magnetic fields in and around sunspots using what was then the recently fabricated spectroheliograph at the Mount Wilson Observatory in 1908. Subsequent observations have shown that sunspots and sunspot groups show a mix of magnetic polarities (both positive and negative or "out of" and "in to" the surface). Examination of the distribution of sunspot magnetic fields points to the existence of another interesting pattern. Hale and collaborator Seth Barnes Nicholson (1891–1963) noticed that all sunspot groups in the northern hemisphere had the same combination of polarities, with all leading spots of one polarity and all following spots of the opposite polarity. In the southern hemisphere, the sign of the leading polarity was

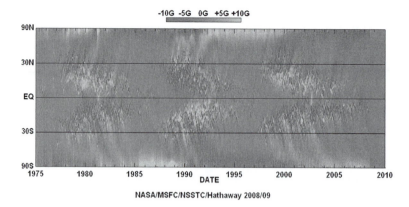

-10G -5G 0G +5G +10G

NASA/MSFC/NSSTC/Hathaway 2008/09

Figure 5.6 The Hale-Nicholson Polarity Law. In each hemisphere all sunspots have the same polarity orientation, which is opposite that of the other hemisphere. Sense of polarity orientation switches with each solar cycle. (Credit: NASA/MSFC/NSSFC Hathaway)

opposite that of the leading polarity in the north (see Figure 5.6). The pattern is reversed with each successive solar cycle. For example, assume that during a given solar cycle all sunspots in the northern hemisphere are found to have a positive polarity leading spot and a negative polarity following spot. In this case, southern hemisphere spots will have a negative leading polarity and a positive following polarity. Eleven years later, spots in the north will have a negative leading polarity, positive following polarity, and so on. This is known as the Hale-Nicholson Polarity Law.

4. Joy's Law

A more subtle pattern was found by the Mount Wilson group in the early part of the twentieth century, namely that the line connecting the centers of the two magnetic polarities of sunspot groups is generally tilted relative to the equator. The tilt is such that the leading spot, in either hemisphere, lies closer to the solar equator than the following spot. The tilt is observed to range from 3 to 11 degrees as the latitude increases up to $+/-$ 30 degrees. This pattern is known as Joy's Law after its principal discoverer, Alfred H. Joy (1882–1973).

5. Hemispheric Helicity Rule

Observations of magnetic fields in sunspots have shown evidence for a large-scale pattern of flows suggesting a coherent spiral structure of the magnetic fields, with the spiral being left-handed, or counterclockwise, in the northern hemisphere and right-handed, or clockwise, in the south. This spiraling property is known as magnetic helicity and is a measure of how twisted the magnetic field is. The first evidence of this hemispheric dependence came, naturally, from observations by George Ellery Hale (1868–1938) in 1927 who noticed a spiral structure in the penumbra of sunspots that spiraled one way in the north and the opposite in the south. This pattern has since been confirmed to be present in the twist of prominences, in S-shaped X-ray structures above sunspots (called sigmoids), and in direct measurements of the twistedness of magnetic fields. This consistent pattern is known as the hemispheric helicity rule.

All the patterns summarized earlier point to a large-scale organization of the production, emergence, and evolution of magnetic field in the Sun. The picture painted by the regularity of the solar cycle, the latitudinal dependence of Spörer's Law, the magnetic relationships implied by the Hale-Nicholson Polarity Law and Joy's Law, and the consistent sense of spiraling in the hemispheric helicity rule is one where a global process inside the Sun determines the local processes observed on the surface and in the solar atmosphere. This led to the development of the solar dynamo model discussed in Chapter 3 and to an understanding of the connection between the solar interior and solar variability.

The photosphere of the Sun has historically been our entry point into the workings of the Sun. Even today, when we can observe the interior and atmosphere in many different wavelengths, the photosphere remains key in developing our understanding of the Sun as a star. The most prominent photospheric features are sunspots, and for almost four hundred years these have opened a window on the global nature of the Sun. Observations of sunspots led us to the notion that the Sun rotates, that it exhibits a cyclical behavior, and that it has a magnetic field—all of which have proved crucial in our continued progress towards finally understanding what drives solar variability.

Starspots

Many cool stars like the Sun are observed to display a level of magnetic activity similar to that of the Sun. The presence of a convective envelope and rotation in such stars fosters the generation of strong magnetic fields resulting in the formation of strong starspots, enhanced X-ray production from the stellar coronae, and large flares. Studying the magnetic activity on other stars allows us to compare the ideas developed for the Sun to a larger class of behavior. Many observational studies have found evidence for spots on other stars, but the properties of these starspots can be very different to those on the Sun.

Temperatures in starspots vary from 3,000 K–4,000 K, depending on the type of star, with the starspot typically being 1000 K–1500 K cooler than the surrounding stellar photosphere. Another major difference is that starspots are inferred to take up 5%-60% of the visible hemisphere.

RECOMMENDED READING

Brown, Michael R., Richard C. Canfield, and Alexei A. Pevtsov, eds. *Magnetic Helicity in Space and Laboratory Plasmas*. Geophysical Monograph. Washington, DC: American Geophysical Union, 1999.

Foukal, Peter V. *Solar Astrophysics.* Berlin: Wiley-VCH, Berlin, 2004.
Ratcliffe, Martin. *Cosmology and the Evolution of the Universe.* Greenwood Guides to the Universe. Westport, CT: Greenwood Press, 2009.

WEB SITES

The Galileo Project: http://galileo.rice.edu.
Mt. Wilson Observatory: http://www.mtwilson.edu.
Sigmoids: http://solar.physics.montana.edu/press/ssu_index.html.
Sunspots: http://www.exploratorium.edu/sunspots.

6

The Ever-changing Sun

Observations of sunspots since the early 1600s have shown that the Sun, far from being static, is dynamic and constantly varying. Everywhere you look on the Sun, it is moving: sunspots appear and disappear, granulation cells come and go, the atmosphere brightens and fades away. On first appearance, this motion seems to be random and chaotic; but observe over long enough timescales and distinctive patterns emerge. In the last chapter, we discussed how almost four hundred years of sunspot observations have allowed us to identify these patterns and to place them in the context of the Sun as a star where the interior, surface, and atmosphere are all interrelated.

The discovery by Heinrich Schwabe in 1843 of a distinctive ten-to-eleven-year cyclic behavior in the variation of the number of sunspots visible on the solar disk stands as one of the most important milestones in the history of solar physics. Schwabe's discovery was based on daily observations he made over a period of seventeen years from 1826 to 1843. Although only regarded as a scientific curiosity in the late-nineteenth century, the Schwabe cycle was eventually recognized as a fundamental property of the Sun. With the subsequent introduction of the Zurich number (see Chapter 5) in 1849 and the development of more detailed sunspot studies, the cycle has been confirmed, and records going back to the early 1600s have been reconstructed—the older records are subject to major uncertainties due in part to the lack of regular observations. The cyclical behavior of the sunspot number indicates a similar behavior in the Sun's magnetic field and the associated activity that it drives. The regular ups and downs of sunspots are accompanied by a corresponding behavior of various active phenomena

Table 6.1. **List of all numbered sunspot cycles**

CYCLE NUMBER NUMBER	DATES OF CYCLE
1	March 1755 – June 1766
2	June 1766 – June 1775
3	June 1775 – September 1784
4	September 1784 – May 1798
5	May 1798 – December 1810
6	December 1810 – May 1823
7	May 1823 – November 1833
8	November 1833 – July 1843
9	July 1843 – December 1855
10	December 1855 – March 1867
11	March 1867 – December 1878
12	December 1878 – March 1890
13	March 1890 – February 1902
14	February 1902 – August 1913
15	August 1913 – August 1923
16	August 1923 – September 1933
17	September 1933 – February 1944
18	February 1944 – April 1954
19	April 1954 – October 1964
20	October 1964 – June 1976
21	June 1976 – September 1986
22	September 1986 – May 1996
23	May 1996 – March 2008 (predicted)

such as solar flares and geomagnetic storms. It is now understood that cyclic variations in activity are an intrinsic property of stars in general.

To keep the record straight and facilitate the discussion of different variations from a "normal" solar cycle, solar physicists introduced a numbering system whereby each successive sunspot cycle occurring since the year 1755 is numbered sequentially. Table 6.1 shows the number and dates of all sunspot cycles recorded since 1755.

Chapter 5 also discussed other patterns evident in the sunspot observations, namely, the Hale-Nicholson Polarity Law, Spörer's Law, Joy's Law, and the Hemispheric Helicity Rule. All of these patterns provide additional insight into the workings of the solar cycle. The polarity law of Hale and Nicholson states that the leading polarity of a sunspot group in a given solar cycle is the same for each group in that hemisphere (north or south), and the leading polarities are different for each hemisphere. Moreover, the sign of the leading polarity switches with the solar cycle. So, in the northern hemisphere sunspot groups will have a positive polarity leading spot in one eleven-year cycle and a negative polarity in the following cycle, and so on. In contrast, the hemispheric helicity rule states that all twisted structures in the northern hemisphere have left-handed twist, while all such structures in the southern hemisphere are right-handed *regardless* of solar cycle. The solar cycle is the large-scale pattern that tells us about the overall behavior, whereas the polarity law and helicity rule provide a level of detail that refines this behavior. As an example, imagine you counted the number of cars traveling on the roads in a big city. You would detect a spike in the number of cars at 8–9 AM and again at 5–6 PM, that is, the morning and evening rush hours (the spans could be longer for certain cities), and this pattern would be repeated every day so you would detect a cyclic activity, specified by the number of cars, with spikes separated by regular eight-hour- and fourteen-hour periods of low activity (ignoring the lack of activity on Saturdays and Sundays). This would be analogous to the solar cycle. If you looked more closely and monitored which direction the cars were moving in general, you would see that in the morning spike the cars were predominantly moving in towards the city, while in the evening spike they would be moving in the opposite direction. This would be analogous to either the helicity or the polarity rule, and would provide additional information that would lead you to the conclusion that the city is a center where people work during the day, but that they predominantly live on the outskirts or in the suburbs.

Terrestrial Impact of Solar Cycles

The cyclical behavior of the Sun has a direct impact on the Earth whether as a direct result of the change in solar radiation or as a consequence of the enhanced array of activity associated with solar maximum. As we will learn in this chapter, the total radiative output of the Sun varies with the number of sunspots, and this has some climatic impact on the Earth. One might assume that the more sunspots present the lower the amount of radiation emitted by the Sun—sunspots are dark, so more sunspots mean less bright photosphere. However, associated with sunspots are bright faculae that are brighter regions of the photosphere surrounding sunspots. The net result is an increase in the overall solar emission at times of solar maximum when the number of sunspots is largest. This change in brightness with the solar cycle has been correlated with the difference in temperature between land and ocean, with this difference tracking the sunspot cycle reasonably well. The variation is very small but it may alter wind patterns, causing marked changes to the local climate, with some evidence for changes in local rainfall totals associated with the sunspot cycle being reported.

A direct relationship between the regular pattern of solar variability and climate on the Earth is made clearer when large variations in the cycle are mimicked by out-of-the-ordinary variations on the Earth. An example of this occurred during the Little Ice Age (see Chapter 11) when a relative lack of sunspots spanning a period of seventy years from 1645 to 1715 coincided with colder and longer winters in the northern hemisphere of the Earth.

Associated with the undulations of the solar sunspot cycle is the corresponding variation in the activity of the Sun, for instance, more solar flares occur at solar maximum. This enhanced activity produces a larger and more diverse range of phenomena at the Earth, with more violent and more frequent geomagnetic storms resulting in increased impact on the human population. For example, enhanced storms can increase radiation dosages to crew and passengers on high altitude polar aircraft flights, can bombard telecommunications and other satellites with energetic ions resulting in disruption or loss of satellite operation, and can, in extreme cases, result in transformer and relay blowouts on electrical grid systems on the ground. Collectively, this array of terrestrial phenomena associated with enhanced activity is known as space weather, which is the topic of a later chapter (Chapter 10).

THE SOLAR CYCLE

While the solar cycle is generally associated with the periodic variations of the number of sunspots, it is of much more fundamental importance to our understanding of the Sun and the physical processes that govern its variability. Since its initial discovery by Schwabe in 1849, evidence for the solar cycle has been found in a diverse array of phenomena both at the Sun and on the Earth. These observations not only yield insight into how the Sun works, but also how it interacts with the Earth and how the Earth responds to changes on the Sun.

Continuous study of sunspots over the last four hundred years has refined our discussions of the solar cycle and allowed us to relate it directly to the physical processes inherent to solar and, indeed, stellar variability. Figure 6.1 shows the sunspot record dating back to the earliest telescopic observations in 1611. The repeatable pattern of the cycle is evident, but a couple of other notable features stand out:

- While the average time between peaks in the sunspot number is around eleven years, it can be as short as eight years and as long as fifteen years.
- Not all cycles are the same, with some cycles exhibiting significantly larger numbers of sunspots than others.
- The larger magnitude cycles tend to display a characteristic shape, with a relatively rapid rise from minimum to maximum followed by a slower decline.

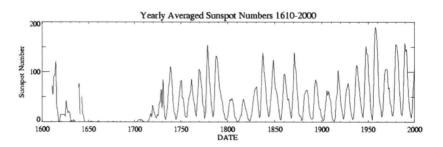

Figure 6.1 Variation of the yearly averaged sunspot numbers from covering four centuries of observations. Courtesy of NASA.

- The cycle amplitudes seem to be modulated on timescales of eighty to one hundred years, with small and large cycles being grouped (minima of this modulation can be seen around 1810 and 1910).
- There is a long period from about 1645 to 1715 with very few sunspots being detected. This is known as the Maunder Minimum.

None of these features have been categorically explained to date. The variation in the duration and shape of each cycle may be no more than random fluctuations about the mean value of 11.1 years, with the rapid rise from the minima indicating that the cycle onset is a rapid response to the reversal of the solar magnetic field (see below), while the decline from maximum is a more gradual transition. The evidence for an eighty-to-one hundred-year modulation (known as the Gleissberg Cycle after the German scientist who first suggested it) is hard to substantiate given the relative short period of time over which we have had reliable sunspot observations. However, proxies for the solar cycle such as carbon-14 (^{14}C) studies, which can go back as far as around AD 1000, show some support for the Gleissberg Cycle, so at present it cannot be ruled out.

The most startling feature in Figure 6.1 is the period of low activity between 1645 and 1715. During this extended period, very few sunspots were detected, despite the fact that sunspots were being observed more or less routinely by astronomers around Europe. Observations from the Observatoire de Paris in France during this period indicate that the average number of sunspots in a given solar cycle fell from around eight hundred to around two hundred, and that in the period from 1690 to 1700 only three sunspots were reported. This period of little or no sunspot activity was first noticed by Gustav Spörer, but made popular by English astronomer Edward Walter Maunder (1851–1928) in 1890. This period is known as the Maunder Minimum. This unusually low level of sunspot activity has been confirmed by other means including the use of carbon-14 measurements in tree trunks (see sidebar). Using proxies of sunspot activity, a number of other extraordinary periods of low or high solar activity have been identified over the last thousand years or so (Table 6.2).

Table 6.2. **Sunspot number extrema**

ACTIVITY	DATES
Oort Minimum	1040–1080
Medieval Maximum[1]	1100–1250
Wolf Minimum	1280–1350
Spörer Minimum	1450–1550
Maunder Minimum	1645–1715
Dalton Minimum	1790–1820

[1] Unlike the multiple periods of activity minima, the medieval maximum represents a 150-year period when the sunspot activity was at a sustained high, which was unsurpassed until the present group of cycles starting around 1950.

The Little Ice Age

The Spörer and Maunder Minima span a period that has come to be known as the Little Ice Age, which denotes a long period of unusually cold weather. There is much debate as to when the Little Ice Age started, with dates ranging from around 1250, when Atlantic pack ice started to grow again, to 1550, when the glaciers started to expand. Whatever the start date, this period spanned a time of atypical climatic behavior—glaciers in the Swiss Alps expanded, rivers, like the Thames in London, froze in the winter, people could walk from Manhattan to Staten Island across the frozen New York harbor, and Iceland's ports were unreachable as ice extended out into the North Atlantic around the island. A more human impact was the increase in floods, famines, and disease brought on by the climatic changes. The Great Famine of 1315–1317 killed millions across Europe as agriculture tried to adapt to the shorter growing seasons.

The cause of the Little Ice Age, while uncertain, is thought to be a combination of decrease solar activity and increased volcanic activity. The former reduces the amount of solar radiation; while the latter increases the reflectivity of the Earth due to the extra dust dispersed across the stratosphere, reducing the amount of solar radiation reaching the Earth's surface. Both of these effects act to reduce the heating of the surface and therefore the overall temperature of the Earth.

Over the last century and a half since the discovery of the solar cycle, a number of studies have shown that a wide range of phenomena exhibit signs of the cyclical variation of solar activity. These indices include solar phenomena such as the sunspot area, the 10.7 cm microwave emission from the Sun, the X-ray brightness of the hot solar atmosphere, total solar irradiance (see Chapter 11), and photospheric magnetic flux and terrestrial phenomena, such as geomagnetic storms, and auroral emission. While the eleven-year periodicity is maintained in these indices, the different physical processes involved mean that the relative variations in amplitude and shape can be significantly different. This is exemplified by comparing the variation of the total solar irradiance over the solar cycle with that of the X-ray emission from the solar atmosphere.

While the variation in the sunspot number over the course of a single solar cycle is dramatic, this in itself has little impact on the Earth. More important is the global output of radiation emitted by the Sun that influences atmospheric physics and chemistry and impacts how much radiation impinges on the Earth's surface. A careful measurement of this emission, called the total solar irradiance, over the last thirty years shows that while it varies in step with the eleven-year sunspot cycle, the amplitude of the variation is of the order of 0.1% (Figure 6.2). The effect of the solar variations on the Earth will be discussed more in Chapter 11.

As a contrast to this small, but significant, variation—a 0.1% change in the Sun's radiative output corresponds to approximately 14 trillion (1.4×10^{13}) Watts of power across the continental United States—Figure 6.3b shows the variation of the atmospheric solar X-ray emission emitted by plasma at 1–2 million Kelvin (see Chapter 7) and the associated variation of the photospheric

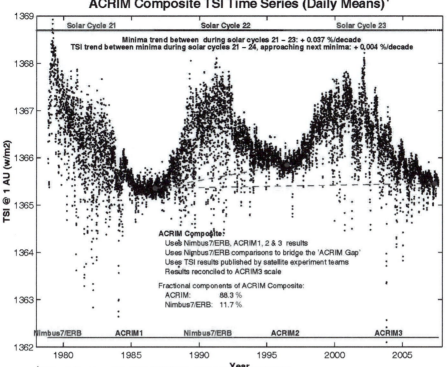

Figure 6.2 Variation of Total Solar Irradiance over two solar cycles. Courtesy of NASA.

magnetic field (Figure 6.3a). The X-ray emission shows a marked variation from the peak of one cycle to the peak of the next, with the total X-ray emission at solar minimum about a factor of 30–50 times smaller than that at maximum. This large X-ray variability is directly correlated with the number and size of the sunspots and their corresponding magnetic field.

Evidence of the solar cycle has also been found on the Earth. In addition, to the tree-ring carbon-14 (^{14}C) signatures, documented reports of auroral activity have proved useful in extending our knowledge of solar activity back as far as the sixth century BC. The northern lights are, predominantly, a high-latitude phenomenon and are most common above the Arctic Circle. Driven by the interaction of the solar wind with the Earth's magnetosphere (see Chapter 8), auroras are visible at almost any time of the year at any phase of the solar cycle. However, increased solar activity results in increased geomagnetic activity and the auroral bands can extend significantly far below the Arctic Circle, reaching as far south as the subtropical zones at times of large solar storms (Chapters 9 and 10). The detection of auroral activity below the Arctic Circle correlates well with the sunspot cycle and gives us a glimpse of the cyclic activity back about 8,000 years. Reliable auroral records from a number of sites in the northern hemisphere date back to around 1450. Earlier than this date, scientists have to rely on

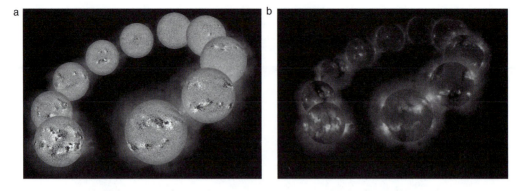

Figure 6.3 Cyclic variation of (a) solar magnetic field and (b) associated X-ray emission. Source: Solar X-ray images are from the *Yohkoh* mission of ISAS, Japan. The X-ray telescope was prepared by the Lockheed-Martin Solar and Astrophysics Laboratory, the National Astronomical Observatory of Japan, and the University of Tokyo, with the support of NASA and ISAS.

written accounts of "celestial events" and their description. Analysis of auroral reports for the 800-year period between 467 BC and AD 333 and the 133-year period from 223 BC to 91 BC (from Titus Livius's *History of Rome*) shows strong evidence for a consistent eleven-year cycle. In addition, radioactive isotopes such as carbon-14, beryllium-10 and chlorine-37 preserved in polar ice can be traced back 100,000 years and has been found to contain evidence for periodicities around 11 (sunspot cycle), 90 (Gleissberg cycle) and 205 (DeVries cycle) years as well as extended periods of global minima.

These proxies are highly valuable as they allow us to extend our knowledge of solar variability further back in time beyond our current multiwavelength monitoring, extending back, at best, three cycles, and beyond the earliest continuous sunspot records, which date back to about 1611. This long-range record of solar activity is useful for shedding light on terrestrial climate discussions such as global warming, recurrent ice ages, and so forth, and will be discussed in more detail in Chapter 11.

Radiocarbon Dating

Carbon-14, also known as radiocarbon is a naturally occurring isotope of regular carbon-12; a carbon-14 atom has six protons and eight neutrons in its nucleus compared to carbon-12, which has six of each. The half-life of ^{14}C is about 5,740 years, and it decays to produce nitrogen—over a span of 5,740 years, half of all of the ^{14}C atoms in a sample will have decayed to nitrogen atoms. Carbon-14 is related to our discussion of the sunspot cycle because it is produced in the upper atmosphere of the Earth when galactic cosmic rays interact with the atmospheric nitrogen. The Earth is constantly bombarded by these cosmic ray particles, but at times of peak solar activity the Sun's magnetic field is stronger and the corresponding interplanetary field gets restructured resulting in a better shielding of the Earth. Thus, as solar activity increases, the cosmic ray flux into the Earth's atmosphere decreases, and vice versa. Lower cosmic ray fluxes at solar maximum means reduced ^{14}C production in the Earth's atmosphere.

The atmospheric ^{14}C gets absorbed by plants, and the information on the concentration levels are preserved. By measuring the ratio of ^{14}C to regular ^{12}C in tree rings, a record of solar activity can be obtained, going back as far as 10,000 years in some cases. While the natural production of ^{14}C tends to smooth out variations on the short eleven-year timescales of the solar cycle, a clear relationship exists between multicycle periods of high, or low, activity and the concentration of ^{14}C detected in tree rings.

In addition to the eleven-year sunspot cycle clearly evident in the four hundred-year record of sunspot observations, arguments have been put forward for the existence of longer cycles in addition to that exhibited by the sunspots. Three additional cycles have been promoted: 88-year Gleissberg cycle, 205-year DeVries cycle, and 2,300-year Hallstatt cycle. The best way to detect a recurring pattern is to have continuous coverage of the phenomenon being considered over a period spanning as many cycles as possible. This ensures that random statistical fluctuations and observational errors have minimal impact on the detection of the recurring pattern. All of the additional solar cycles proposed require solar activity records going back further than the recorded sunspot observations. This requires the use of activity proxies, some of which have been discussed in the main text. These proxies, such as the ^{14}C concentrations in tree rings, have been shown to provide a good match with the sunspot data over the last two hundred to four hundred years and so can be reliably used as a measure of solar activity in the past. Proxies such as the ^{14}C tree ring concentrations, ^{10}Be concentrations in polar ice, and auroral records all show strong evidence for an eighty-five- to ninety-year fluctuation supporting the case for the Gleissberg cycle first proposed by German scientist Wolfgang Gleissberg (1903–1986) in 1958. Somewhat weaker, but suggestive, evidence exists in the radioisotope proxies for a cycle of order 200 years first pointed out by Dutch astronomer Hessel de Vries (1916–1959) in 1958. The time between the Spörer and Maunder Grand Minima (see Table 6.1) in the sunspot record is approximately two hundred years and thought to be part of the DeVries cycle. Analysis of the ^{14}C data shows evidence for many more cycles over the 10,000-year record. For example, the existence of a millennial-scale cycle of roughly 2,300 years, named the Hallstatt cycle, has been speculated based on an association with two cold spells, the first of which occurred during the Hallstattzeit (1200–500 BC), a Bronze Age and early-Iron Age period named after the region of Austria in which a large burial ground was discovered.

So far we have discussed the solar cycle as a single-dimensional time series of a repeatable pattern. However, a closer look at the behavior of sunspots as they appear on the solar disk illuminates the presence of another pattern, namely, that at the beginning of a solar cycle sunspots appear at solar latitudes of around 20–40 degrees, while at the end of the cycle they appear near the equator (latitude 0°). This behavior for the last nine cycles is shown in Figure 5.5, and, for obvious reasons, this sequence is known as the butterfly diagram. The butterfly diagram indicates the latitudinal variation of sunspot emergence over the course of a solar cycle. This latitudinal drift is sometimes referred to as Spörer's Law.

The solar cycle can now be viewed as a combination of when and where the sunspot emerges. A simple count of sunspots takes no account of the Hale-Nicholson polarity law, which indicates that spots in successive cycles have opposite magnetic orientations (in the same hemisphere). A key point here is the fact that the butterfly diagram shows us that the wings of the butterfly in one cycle overlap in time with the nose of the butterfly in the preceding cycle, with the first spots of the new cycle appearing at a latitude of 25°–30° and the last spots of the old cycle appearing at latitudes less than 20°. This overlap has been known for more than 130 years and indicates that the cycles are not completely sequential. New cycle spots can be observed at the same time as old cycle spots, differentiated by their magnetic polarity. The time between the appearance of the first and last spots of a given cycle have been found to range from 12 to 14.2 years, that is, significantly longer than the 11.1-year average time between successive solar minima. The latitudinal drift of the appearance of sunspots as a solar cycle progresses provides important clues to the generation of magnetic field in the interior of the Sun.

THE MAGNETIC CYCLE

It was noted in the last chapter that sunspots were intrinsically linked with the solar magnetic field. The large-scale patterns exhibited by sunspots, the cyclic behavior, the latitudinal pattern of emergence (Spörer's Law), the polarity rule (Hale-Nicholson Polarity Law), and the equatorward tilt (Joy's Law) all provide crucial information on the generation and evolution of the magnetic field on the Sun. Any model purporting to explain the solar magnetic field has to reproduce the large-scale patterns exhibited by sunspots.

In terms of the cyclic behavior of the magnetic field, the following three main characteristics are important to bear in mind:

- The Hale-Nicholson Polarity Law states that all leading spots in a given solar cycle in a given hemisphere have the same polarity, while all leading spots in the other hemisphere have the opposite polarity *and that the leading polarity in each hemisphere changes with each new cycle.*
- The magnetic field undergoes a dramatic change over the solar cycle in concert with the waxing and waning of the sunspots.
- The polarity of the magnetic field at the solar poles reverses with each sunspot cycle.

The polarity reversal of the magnetic field is mediated by a drift of an opposite polarity magnetic field towards both the north and south poles of the Sun. This is a result of the diffusion of sunspot regions as they evolve.

Joy's Law, which states that the leading polarity lies closer to the equator than the following polarity, reflects the separation in latitude of the sunspot polarities, which grows as the sunspot region gets older. One polarity drifts equatorward to interact with the opposite polarity drifting from the opposite hemisphere, to cancel and nullify the field. The other polarity drifts poleward to interact with opposite field from the previous cycle. Eventually, over a period of one to two years, the field at the poles is replaced with opposite polarity field. The polar reversal of field occurs near sunspot maximum and is about half a cycle out of phase with the polarity switch of the sunspots (Hale-Nicholson Polarity Law).

The behavior of the magnetic field introduces another representation of the solar cycle, which some solar physicists argue is actually the more important representation. The solar magnetic cycle, or Hale cycle, is the time over which the large-scale magnetic configuration of the Sun repeats. The Hale cycle is twenty-two years long, or twice the sunspot cycle—this is the time it takes for the polar fields to repeat.

As described in detail in Chapter 3, the relationship between the sunspot cycle and the magnetic Hale cycle began to take shape after the development of the magnetograph by Harold and Horace Babcock in the 1940s. As the connection emerged with improved data, Horace Babcock (1912–2003) developed a model, known as the Babcock model, which describes the twenty-two-year magnetic cycle as a series of stages culminating in polarity reversal of the large-scale magnetic field, after which the process repeats. The Babcock model was described in detail in Chapter 3. The importance of this model was that it provided a simple and direct means by which the low-latitude sunspot fields were connected to the high-latitude polar fields through their evolution, and that the overall behavior of the surface fields was directly related to the subsurface generation of the field and driven by a series of processes like differential rotation, meridional flows, and magnetic buoyancy. The Babcock model successfully explained the Spörer, Hale-Nicholson, and Joy Laws and paved the way for more complicated solar dynamo modeling (Chapters 3 and 4). In the Babcock model, the large-scale field reversal is a result of the poleward migration of the following polarity spots and the equatorward migration of the leading polarity spots. These fields interact with their opposite counterparts, effectively reversing the field.

KICKING OFF THE CYCLE

Our basic description of a solar cycle starts with the Sun at solar minimum when the number of spots is low. At the beginning of a new cycle, the large-scale field is relatively simple and is basically in the form of a magnetic dipole with the north-south polarity axis aligned with the Sun's rotation axis. This is evident in the observations of the solar corona (Chapter 7), which generally has a continuous belt of emission (the helmet streamer belt) and regions of negligible emission at each of the poles (coronal holes).

The tell-tale sign that the new cycle is beginning is the appearance at mid-latitudes (around 30°) of a sunspot region that violates the Hale Nicholson Polarity Law for the cycle just finishing. Despite the low number of sunspots present at solar minimum, it can be difficult to detect the new cycle spots, which tend to be small and short-lived. As the cycle develops, the Hale-Nicholson polarity pattern emerges as more new spots develop. Old and new cycle spots are observed to overlap by two-three years. As the new cycle activity begins to dominate the old cycle activity, with more new cycle regions of reversed polarity emerging at latitudes around 30°, the number of "new" old-cycle spots drops and any existing ones disperse with their following polarities drifting polewards, and the leading polarities drifting equatorward. Most of the new magnetic activity is characterized by small, short-lived sunspot regions, but within a year or so of the new cycle starting, the magnetic activity rapidly increases with most of the magnetic flux residing in strong field sunspot regions.

At this point, and over the next three to four years, the emergence of a large amount of strong magnetic flux results in the large-scale magnetic field departing significantly from the simple dipole field of solar minimum to a more complex configuration. The polar fields have reversed their polarity and now reflect the half-cycle configuration of the Hale cycle. The coronal helmet streamers and coronal holes are no longer confined to equatorial

Table 6.3. **Characteristics of solar cycle extrema**

SOLAR CYCLE CHARACTERISTICS	
Solar Minimum	Few sunspots on disk
	Sunspots at low latitudes
	Little or no strong magnetic field
	Weak atmospheric emission
	Large-scale dipole field with axis aligned to Sun's rotation axis
	Continuous equatorial streamer belt
	Polar coronal holes
	Few flares and coronal mass ejections
Solar Maximum	Many sunspots on disk
	Sunspots at mid-latitudes
	Strong magnetic field concentrations
	Enhanced coronal emission
	Multipolar complex large-scale field
	Distributed helmet streamer system
	Equatorial extensions on polar coronal holes
	Many flares and coronal mass ejections

and polar regions, respectively, but can occur anywhere on the Sun. These excursions of the helmet streamers and equatorial extensions of coronal holes significantly modify the interplanetary magnetic field structure and the solar wind that have strong repercussions for space weather, which will be discussed in subsequent chapters.

About seven years after the start of new cycle activity, the number of new sunspots starts to decline, and when new sunspots appear they do so at lower latitudes as we head to the next solar minimum. As the cycle declines, the reduction in the amount of strong magnetic field reduces the complexity of the large-scale field and the dipole is restored, although with the polarities reversed and at a large tilt angle to the Sun's rotation axis. It takes a further year or two for the dominant dipole field to grow in strength sufficiently to align its axis with that of solar rotation. We are now at the beginning of our story and the whole process begins again.

Active Longitudes

Sunspots do not emerge at random over the solar disk. As well as the active latitudes, which vary equatorward over the course of a solar cycle, as seen in the butterfly diagram, sunspots also tend to prefer specific longitudes, or active longitudes. The phrase "active longitude" is somewhat misleading as it suggests that sunspots emerge all along the longitudinal line and does not take account of the localized latitudinal emergence. Consequently, the more suggestive *active nests* have been adopted. Observations spanning several rotations show repeated emergence of sunspots in the same localized area sometime before the existing sunspot regions have dispersed. Something like 35%–45% of all sunspot groups appear in compact active nests, with the active nests being more common near solar maximum. It has been suggested that the existence of active longitudes is a result of the dynamo process at the base of the convection zone near the tachocline. Bulges in the toroidal field at these depths create an instability that is longitudinally dependent and results in a larger probability of magnetic field emerging at these locations.

PREDICTING THE SOLAR CYCLE

Understanding the solar cycle and being able to predict the expected level of activity in advance is important to us here on Earth as it has an impact on our daily lives, especially now, in the twenty-first century, when we rely so much on advanced technology, much of it based on resources in space. We use a wide range of low-Earth orbiting satellites for telecommunications, weather monitoring, TV and radio broadcasting, and navigation. Increased solar activity can severely interfere with or even damage these resources with major impacts on the surface—imagine not knowing when a hurricane was going to strike your town until it was blowing the roof off of your house. Other effects of increased solar activity include impacting electrical power grids on the ground, increasing the erosion of oil pipelines, and causing the Earth's atmosphere to expand, thereby increasing the atmospheric drag on satellites. Advanced knowledge of the expected magnitude, shape, and duration of the next solar cycle is, therefore, crucial for logistical planning and disaster preparedness.

Given the importance of solar activity, which is closely tied to the ups and downs of the solar cycle, there is much interest in being able to predict the future evolution of the solar cycle based either upon "past performance" or on detailed modeling. The former approach uses patterns in a number of different indicators to see if there is any relationship between these patterns and the strength or duration of the solar cycle. There is very little real physics involved in this approach as it relies more on statistical inferences, that is, "the last time the Sun did this the following cycle was unusually strong." The latter approach is based on recent developments in understanding the underlying physics of solar activity, namely the solar dynamo (see Chapter 3). The cyclic behavior of the large-scale solar magnetic field is thought to be due to a dynamo process operating inside the Sun, and models of this process have successfully explained many of the patterns of activity outlined in Chapter 5.

If solar activity followed a simple regularly periodic cycle with a fixed amplitude, then predicting all future behavior would be relatively straightforward. However, as we have already shown, measurements of solar activity using direct sunspot observations over the last four hundred years or by using proxies for this activity indicate that the variations in the magnetic activity depart frequently from a strictly periodic behavior. Not only do the cycle magnitudes vary significantly, but also the actual durations of the cycles and time between successive minima can be very different. The solar cycle is strongly modulated so long-term periodicities, like the Gleissberg cycle, are superimposed on the eleven-year Schwabe cycle. Extreme examples of this modulation include the various grand minima, for example, Spörer and Maunder Minima, and the medieval maximum. The proxy signatures from radio isotopes ^{14}C and ^{10}Be suggest that this modulation of the solar cycle is a characteristic feature spanning the last 10,000 to 20,000 years. Notwithstanding this modulation, which can severely hamper the ability to predict future activity at the levels required to be useful, several attempts have been made to bring together all of our current knowledge of solar magnetic behavior to make accurate predictions for solar activity at least a few years in advance.

In the statistical approach, a wide variety of solar and geomagnetic activity indicators are used to identify trends and pointers in the preceding cycle to predict the characteristics of the forthcoming cycle. Such indicators include the geomagnetic indices reflecting global levels of geomagnetic activity, which respond to the changing conditions in the solar wind, the number and timing of high-latitude spots, the strength of the Sun's polar fields, the radio flux of the Sun at 10.7 cm wavelength (a proxy for solar activity), and the number of days on which the Earth's magnetosphere was in a "disturbed" state (generally defined from enhanced geomagnetic values).

These methods have proved reasonably successful, but the Sun always seems capable of springing a few surprises. In recent solar cycles, Cycles 19 through 22 (see Table 6.1), the trend was towards larger amplitudes. Cycle 19 was the largest on record with a smoothed sunspot number maximum of 201 occurring in March 1958, while Cycle 22 was the third largest (maximum sunspot number of 159 in July 1989). In addition, both Cycles 21 and 22 exhibited historically high levels of geomagnetic activity. All this pointed to a prediction of a large Cycle 23 with a peak sunspot number in the range 100–240 (depending upon the method used) occurring in March 2000 with a range of ±14 months. The actual maximum of Cycle 23 was double peaked with a first peak occurring in April 2000 having a smoothed sunspot number of ~121 and a second peak occurring in November 2001 with a smoothed sunspot number of ~115. Most of the methods used to predict the peak of Cycle 23 predicted much higher sunspot counts. Only those methods that employ solar precursor data, that is, take account of the global magnetic state of the Sun, predicted values in the correct range. None of the methods predicted a double peak. Having reached the minimum of Cycle 23, predictions for Cycle 24 have already been proposed, although, as usual, the predictions do not agree with one another. Some predictions argue for a much weaker cycle than Cycle 23, some argue for a similar level cycle, and others argue for a much stronger cycle. The latest combination of predictions suggests that Cycle 23 will end in March 2008 ±6 months with Cycle 24 starting at about the same time. Solar minimum was reached in March of 2008 signifying a change in the cycle and confirming the prediction. The peak of Cycle 24 is predicted to occur either in October 2011 with a maximum sunspot number of 140 ± 20 or in August 2012 with a maximum sunspot number of 90 ± 10 (see Figure 6.4). An average cycle has a maximum sunspot number around 114. The split in times and peak values is the result of a division in the panel of scientists who collate the various predictions and develop a consensus estimate for the next solar cycle. There was a clear split on this panel between a higher-than-average cycle occurring early and a smaller-than-average cycle occurring late.

Successful predictions of future solar cycles have also been made using sophisticated models of the solar dynamo that governs the physics of magnetic field generation and evolution. The large-scale solar dynamo used involves three basic processes (see Chapter 3), the generation of toroidal field from the shearing of a preexisting poloidal field by differential rotation (the Ω effect), the regeneration of new oppositely directed poloidal fields by lifting and twisting the toroidal flux tube (the α effect), and the inclusion of the transport of flux by meridional circulation. Combining all of these three factors with dynamo modeling is proving quite successful at predicting the gross features of the solar cycle. Dynamo models used to address the cycle prediction problem must explain the following observed phenomena:

activity, while commonly observed on other stars, is not observed on all of them. The lack of cyclic activity for certain stars can tell us much about the whole phenomenon of cyclic variability.

A large variety of stars, including the Sun, have a number of significant phenomena in common. Our understanding of the Sun has led to observational searches for solar-like phenomena on other stars, and we now know that, like the Sun, stars can have convection zones, granulation, magnetic fields, starspots, flares, and cyclic activity. For example, magnetic field strengths of order 1,000–2,000 Gauss covering 20%–80% of the visible area of the star have been detected. These field strengths are comparable to what we find in sunspots, but on the Sun these regions of strong field take up less than 1% of the visible surface. This implies that a much higher level of activity can be supported on these more energetic stars. In this context, the Sun is a relatively low-activity star.

Because of the sheer distance to the stars and the quality and size of instrumentation required to observe them, few indicators of activity are sensitive enough to cyclic variations to generate a strong enough signal to be observed from the Earth. The indicators that can be used include some EUV lines, the Hα optical line, the infrared He 1083 nm line, and, most commonly, the optical emission from the Calcium II (CaII) H and K lines. The CaII H and K lines, at wavelengths of 393.4 nm and 396.8 nm, respectively, are strong absorption lines in singly ionized calcium atoms generally produced at plasma temperatures indicative of stellar chromospheres. The Ca H and K emission from the Sun clearly shows the eleven-year activity cycle with a cycle amplitude (between minimum and maximum) of order 10%.

Long-term, ground-based spectroscopic monitoring of the Ca II H and K activity proxies of approximately one hundred main sequence stars from the Mount Wilson Observatory, known as the HK Project, has demonstrated the existence of different types of stellar variability.

The accumulated results from the forty-year history of the HK Project indicate that more than half (60%) of the stars in the sample exhibited cycles, with the interesting result that very few cycles less than six to seven years long were detected, suggesting a lower limit to the operating timescale of the cycle-generating mechanism. The cycle amplitude of these stars ranges from a solar value of order 10% to an extreme of 35%. When stars are grouped by their spectral type, it is found that older stars tend to have more regular cycles while young, highly active stars tend to be more irregular.

Overall, the HK Project has demonstrated that many stars, other than the Sun, exhibit cyclical activity variations, but that some show no sign of a regular variability at all. The combination of stars with and without cycles provides a database from which it should be possible to identify the key stellar parameters (mass, luminosity, age, rotation rate) required to drive cyclic activity, and so provide a window on to the key physical processes responsible for this variability. Results indicate that both the rotation rate and the depth of the stellar convection zone are the principal stellar quantities for determining the amplitude and duration of the stellar cycle. Rapidly rotating active stars exhibit large amplitude cycles, although they also tend to be quite irregular, with a suggestion that solar-like activity kicks in when the rotation period exceeds twenty days (mean solar rotation rate is around twenty-five days). Stars without convection zones (hot O and B stars) tend not to show cyclic variability, while stars with convective zones show a rotation-dependent variability suggesting that the presence of a significant convection zone is crucial to determining the form of the stellar activity.

The observed association between magnetic activity and rotation in stars gives a strong clue to the physical process driving the activity. Regardless of stellar age and spectral type, solar-like stars show a remarkable consistency in the relationship between their chromospheric emission, as determined by the Ca K observations, and their rotation rate, indicating that a common mechanism is responsible for determining the level of activity. This relationship can be understood in terms of the role played by differential rotation in the dynamo process discussed in Chapter 3. The understanding of the dynamo process on the Sun can be coupled to our understanding of stellar activity relationships to refine the dynamo models and enhance our understanding of the generation of magnetic fields in the astrophysical bodies. The clear connection between magnetic activity and the heating of the solar atmosphere (Figure 6.3 and Chapter 7) means that a study of stellar variability may prove useful in constraining the mechanisms responsible for heating the atmospheres of stars and ultimately improve our understanding of a star's impact on its surrounding planets.

RECOMMENDED READING

Fagan, Brian M. *The Little Ice Age: How Climate Made History, 1300–1850.* New York: Basic Books, 2000.

Wilson, Peter R. *Solar and Stellar Activity Cycles.* Cambridge: Cambridge University Press, 2005.

WEB SITES

The HK Project: http://www.mtwilson.edu/hk.

Predicting Cycle 24: http://www.swpc.noaa.gov/SolarCycle/SC24.

Solar Cycle model: http://www.ucar.edu/news/releases/2006/sunspot.shtml.

Sunspot Cycle: http://solarscience.msfc.nasa.gov/SunspotCycle.shtml.

7

The Many Faces of the
Solar Atmosphere

It seems as strange to talk about the Sun having an atmosphere as it did to talk about it having a surface. However, the properties of the outer portions of the Sun are such that there is a clear division in both temperature and density between the photosphere (surface) and the material that lies above it. Effectively, the Sun's atmosphere extends throughout the solar system, only "officially" stopping at the heliopause where the pressure of the Sun's atmosphere is balanced by that of the interstellar medium. The whole solar system is bathed in the atmosphere of the Sun, and this can have interesting consequences, which are the subject of later chapters.

A simple, although incorrect, picture of the solar atmosphere describes the different layers as one moves upward from the surface. In this picture, the temperature increases from the temperature minimum region of the photosphere as the density rapidly falls off, defining three primary regions: the chromosphere, the transition region, and the corona. These regions will be discussed in detail in this chapter, with a particular emphasis on the corona, which provides the main interaction with the planets of the solar system.

In this layering description, each region of the solar atmosphere represents a spherical shell that has a lower and an upper boundary at a height specified by the temperature variation. Table 7.1 shows the approximate height, temperature, and density ranges of the three atmospheric regions, where the top of the photosphere is assumed to have height zero. The heights given in the table can be misleading since spicules, which are

Table 7.1. **The characteristics of the solar atmosphere**

	TEMPERATURE (K)	DENSITY (m^{-3})	HEIGHT RANGE (km)	PROMINENT FEATURES
chromosphere	$4,500 - 2.5 \times 10^4$	$2 \times 10^{16} - 2 \times 10^{17}$	$500 - 2300$	Prominences Spicules Flares
transition region	$2.5 \times 10^4 - 10^6$	$10^{14} - 2 \times 10^{16}$	$2300 - 2800$	Postflare loops
corona	$(1-5) \times 10^6$	$10^{14} - 10^{15}$	>2500	Hot loops Streamers Flares Coronal holes

well-known chromospheric phenomena, can reach heights of over 5,000 km when viewed at the solar limb, and prominences can be seen to heights of several tens of thousands of kilometers. However, the values for temperature, density, and height range are average values determined from a model of the solar atmosphere and are representative of the major differences between the three atmospheric regimes.

The division of the solar atmosphere into layers is artificial as it takes no account of the detailed structure of the atmosphere where plumes of chromospheric material reach up into coronal heights, and low-lying plasma can be heated to coronal temperatures. This detailed structure of both the chromosphere and corona provides insight into the transfer of energy and the conversion of that energy into heat and motion (Figure 7.1). Strictly speaking, the designation of a volume of plasma in the solar atmosphere as chromospheric, transition region, or coronal plasma only requires knowledge of its temperature and not its height.

One of the most startling aspects of the solar atmosphere is that it is extremely hot, much hotter than the photosphere. Typical chromospheric temperatures are in the range 6,000 K to 20,000 K, while coronal temperatures vary from about 1,000,000 K (1 MK or 1 megaKelvin) in the quiet Sun to 3–5 MK above sunspots and can reach 50–100 MK in solar flares (Chapter 9). What makes this surprising is that all the energy emitted by the Sun is generated in the nuclear fusion processes of the solar core. Throughout the solar interior, the temperature falls off with distance out to the photosphere, just as you would expect—it gets colder the further you stand from the campfire. Then, things get strange. As you go higher in the solar atmosphere, that is, further away from the solar center, the temperature starts to climb. In the almost seventy years that solar physicists have known how hot the solar corona is, there is still no clear understanding of

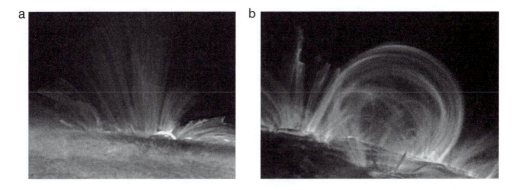

Figure 7.1 High-resolution observations of (a) the solar chromosphere from *Hinode* and (b) the solar corona from TRACE.

how the atmosphere gets so hot. This is known as the *coronal heating problem.*

The solar atmosphere is a complex region encompassing a wide range of temperatures, densities, dynamics, and physical processes. It is pervaded by magnetic field that provides much of the energy needed for the phenomena observed and a conduit by which this energy is distributed throughout the heliosphere. This chapter tries to give a flavor of some of this complexity and how the different parts of the solar atmosphere are interrelated.

OBSERVING THE SOLAR ATMOSPHERE

Observations of the outer atmosphere of the Sun have been available since antiquity, although it has only been acknowledged as an atmosphere in the last two centuries. Accounts date back to more than 2500 BC in Chinese records and continue throughout all of the world's civilizations to the present day. Eclipse observations, recorded by all ancient cultures, Chinese, Babylonian, Chaldean, Indian, provide a spectacular display of this atmosphere, earning the name *corona* for its crown-like appearance (Figure 7.2). Ancient eclipses were seen as prophetic omens, and astronomers who could predict their occurrence were granted great favor by the emperors, kings, and rulers whom they served. Different cultures had varying ways of explaining the temporary disappearance of the Sun, including the ancient Chinese belief of a sun-eating dragon that had to be scared away by making as much noise as possible. Regular scientific observations of solar eclipses began in the early eighteenth century, with photographic records beginning in 1851 during the Scandinavian eclipse that crossed Norway and Sweden on July 28 of that year.

The popular attraction of eclipses remains strong today, but their scientific value has steadily declined since the invention of the coronagraph by French astronomer Bernard Lyot (1897–1952) in 1930. The Lyot

Figure 7.2 1991 eclipse image courtesy Rhodes College, Memphis, Tennessee, and High Altitude Observatory (HAO), National Center for Atmospheric Research (NCAR), Boulder, Colorado. NCAR is sponsored by the National Science Foundation.

coronagraph essentially creates a solar eclipse by using a carefully tailored circular aluminum disk to block out the main disk of the Sun. The startling structure of the solar corona is then available to the astronomer on a daily basis for scientific study. Since Lyot's original invention, we now have coronagraphs in a number of observatories around the globe and in space, with the most prominent being the Mark IV coronagraph of the Mauna Loa Solar Observatory on the Big Island of Hawai'i, the LASCO (Large Angle Spectrometric and Coronagraph experiment) coronagraphs on board the SOHO spacecraft and the STEREO (Solar Terrestrial Relations Observatory) spacecraft containing a pair of coronagraphs, one on each of the two STEREO spacecraft, as part of the SECCHI (Sun Earth Connection Coronal and Heliospheric Investigation) instrument suite. The STEREO coronagraphs will view the solar corona simultaneously from two separate locations along Earth's orbit, providing a 3D view of the corona. Despite the cumbersome nature of the acronyms, coronagraphic observations of the Sun have proved crucial in understanding the large-scale structure of the corona and the explosive nature that it occasionally displays (see Chapter 9).

The corona is so hot (1–3 million Kelvin) that the radiation it produces is too energetic to be seen in the optical part of the spectrum, but rather emits in the extreme ultraviolet and X-ray ranges. An additional consequence of this hot temperature is that the atoms that make up the solar

Temperature and Wavelength

The higher the temperature of a plasma, the higher the energy of the photons it emits. The energy of the photon is defined by the temperature of the emitting medium, and the properties of light relate this energy to the wavelength or frequency of the photon. Simply speaking, we have peak wavelength = $2.9 \times 10^6 \, T^{-1}$, so the greater the temperature, T, the shorter the wavelength of peak emission. This is known as Wien's Law. Thus, a photon representing a 3 million degree corona has a wavelength of 1 nm, which is well in the X-ray range of the electromagnetic spectrum. A photon from the 5,800 degree photosphere, on the other hand, has a wavelength of 500 nm at the green end of the visible range.

corona are fully ionized: the temperature is so hot that all of the electrons get "kicked" off the atoms. It is the presence of the free electrons that serves to "illuminate" the corona at optical wavelengths by a process known as *Thomson scattering* (see sidebar). Effectively, we see the light from the solar surface that gets scattered by the electrons towards the Earth. The scattering process is sensitive to the number of electrons present in a given location, allowing us to see the high-density structures that delineate the solar corona. Coronagraph observations have proven invaluable in determining the structure, evolution, and variability of the solar corona, and the underlying physical processes that drive it. Coronal mass ejections (CMEs), which provide a crucial link between the activity of the Sun and the impact of that activity on the Earth (Chapter 9), are an example of a major discovery resulting from coronagraph observations.

Thomson Scattering

Thomson scattering is the name given to the scattering of electromagnetic radiation by particles, such as electrons, named after the British physicist J. J. Thomson (1856–1940), who first explained the process. The electromagnetic radiation, which for our purposes can be regarded as the light emanating from the solar surface, can be regarded as a wave that interacts with the particles, for example, the free electrons making up the solar corona. This interaction accelerates the particle, which in turn reradiates its excess energy in a different direction from that of the incident wave (see Figure 7.3). From the perspective of an outside observer, the original wave (or light) is

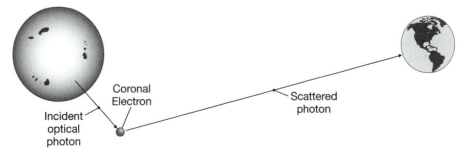

Figure 7.3 Diagram showing mechanism of Thompson scattering. Illustration by Jeff Dixon.

scattered. In the case of a solar eclipse, the direct light from the solar surface is blocked by the moon, but some of the light from the surface traveling perpendicular to the Earth direction gets scattered by the coronal electrons towards the Earth, allowing us to "see" the corona. The solar K-corona is the result of the Thomson scattering of solar radiation from solar coronal electrons.

Spectroscopic observational techniques of the Sun (see Chapter 2) were developed specifically for use during solar eclipses, and this ultimately led to the identification of the solar corona as an integral part of the Sun not visible against the brightness of the solar disk. The spectroscopic measurements ultimately led to the major surprise, via the detection of emission from what were later discovered to be highly ionized iron ions, that the corona was much hotter than the solar photosphere (see below). An earlier, but just as startling, discovery was of an entirely new element that had as yet never been seen on the Earth. This element was helium (see sidebar). Spectroscopic observations also played a crucial role in the discovery of the different temperature regimes of the Sun and, in particular, the detection of the chromosphere and transition region.

Discovery of Helium

As discussed in detail in Chapter 2, the advent of spectroscopic observations of the Sun initiated by the work of Fraunhofer, Kirchhoff, and Bunsen in the nineteenth century led to many new discoveries about the composition of the Sun and its basic properties, like temperature and density. Many of these discoveries relied on the identification of similar spectroscopic signatures in the laboratory on Earth. For instance, we would not know that the two strong yellow lines, observed at wavelengths around 589 nanometers, were emitted by the element sodium if not for their detection in the flames of burning liquids containing sea salt, sodium chloride—these lines were first detected, but not identified, by Scottish scientist Thomas Melville (1726-1753) in 1752.

However, spectroscopic observations of the Sun also led to some interesting surprises. In 1868, English physicist Sir Norman Lockyer (1836–1920) detected a bright yellow line in his spectroscopic observations of a solar prominence (this line was observed in the same year by French astronomer Pierre Janssen). This yellow line appeared at a wavelength of 587.56 nanometers, where none of the known elements had emission lines. A subsequent search by Lockyer and Janssen also failed to produce a laboratory equivalent of the solar observations. Lockyer concluded that the line he observed was the fingerprint of an unknown element, unique to the Sun, which he named *helium* after the Greek word for Sun, *helios*. Despite being ridiculed by his peers, Lockyer was proved correct twenty-seven years after his initial observations when Scottish chemist Sir William Ramsay (1852–1916) discovered a mysterious yellow line at 587.5 nanometers in the spectrum of a mineral of uranium called cleveite. This line was identified as the self-same line observed on the Sun by Lockyer.

While we know helium as the gas that makes party balloons float and gives us squeaky Donald Duck voices when inhaled, it is the second most abundant element in the universe, after hydrogen. It is tasteless, odorless, and colorless, which is why it was so difficult to detect in the first place. Helium also has the distinction of being the only element ever to have been discovered on the Sun before it was discovered on the Earth.

The chromosphere was detected as a rose-colored layer of gas surrounding the visible disk of the Sun that flashed briefly near the end of a solar eclipse. Spectroscopic studies have now characterized some of the basic properties of the solar chromosphere, but even now, some 150 years after its discovery, the chromosphere remains largely a mystery. The chromosphere is difficult to study in detail since the processes that produce the observed radiation are far more complex than in the solar corona. The range of density and temperature across the chromosphere means that while a significant fraction of the material is ionized, much of it is in the form of neutral atoms. The mixture of charged and neutral material together with the high density significantly complicates the emission and absorption of radiation, making it difficult to accurately infer the chromospheric conditions from what finally escapes to be detected in our spectrometers. The chromosphere is also very dynamic, with the plasma racing back and forth along and across the magnetic field. The spectroscopic observations of the chromosphere have often been described as resembling a "prairie on fire" since vast plasma jets and spicules are seen to erupt from the chromosphere (Figure 7.1a). An important chromospheric feature is the solar prominence, observations of which date back to 1733. Prominences are extremely important in modern-day solar physics, and will be discussed in more detail at the end of this chapter. Much of the energy transfer throughout the chromosphere is a result of the generation and dissipation of wave motions, which further complicates the dynamical picture. Despite the complexity of the physics in the chromosphere, progress is being made on understanding this extremely important region of the solar atmosphere through which all of the energy required by the corona must pass.

The sharp increase in temperature from the chromosphere to the corona is mediated by the *transition region,* a very narrow region (100–500 km across) in which the temperature rises from the order of 20,000 K at the base to one million degrees at the top. The short distance over which this transition occurs means that the temperature gradients (how quickly the temperature changes with distance) are extremely high, leading to some rapid changes in the properties and, therefore, the physics of that region of the solar atmosphere. The small scale and the large gradients make the transition region extremely difficult to observe. Modern observations rely on specially designed telescopes that target the temperatures associated with the transition region. At extreme ultraviolet (EUV) wavelengths, which can only be observed from space, plasma at transition region temperatures can be seen as elongated structures, or magnetic loops, and as a moss-like expanse permeating active regions.

In addition to the rapid variation in temperature, there is an associated variation in density. The combination of these changes results in a transition from a mostly opaque, partially ionized chromosphere, where most of the emission occurs in the infrared, visible, and ultraviolet parts of the electromagnetic spectrum, to a mostly transparent, fully ionized atmosphere where low-wavelength (extreme ultraviolet and X-ray) emission dominates.

STRUCTURE OF THE SOLAR ATMOSPHERE

The large-scale corona, as observed in solar eclipses, has a great deal of structure, most notably the large pointed helmet streamers. However, this structure is embedded in a continuous distribution of coronal emission. Since the eclipse observations of the nineteenth century, we have known that the solar corona has four components: the E-, K-, F-, and T-corona. The K- and F-corona are the most prominent components. The E-corona represents the emission line component of the corona detected by the first spectroscopic measurements of the solar atmosphere, but not understood until almost seventy years later. The K-corona is probably the main component of the corona (the K stands for *Kontinuum*, named by its German discoverers), extending several solar radii into space. The K-corona is a result of the scattering of light from the solar surface by high-velocity free electrons in the corona. The free electrons are a "by-product" of the ionization process of coronal elements. This scattering is known as Thomson scattering (see sidebar) and serves to smear out the strong line spectrum of the solar surface (see Chapter 5) due to the motions of the scattering electrons. The F-corona (*Fraunhofer corona*) is a result of the scattering of photospheric light by slow-moving particles of dust and extends far out into the interplanetary medium. Near the Earth the F-corona is observed as a faint emission lying along the ecliptic plane known as zodiacal light. Because the dust particles are much more slowly moving than the electrons in the K-corona, the Fraunhofer lines of the solar spectrum essentially maintain and are seen as prominent structures in F-corona spectra. It should be clear from this description that the F-corona is not part of the solar atmosphere as it is comprised of remnant dust and light from the solar surface. The T-corona is the name given to the thermal emission from the interplanetary dust.

The atmosphere of the Sun is highly structured, and a description of it is probably best served by describing the various features that make up the various components of what we call the solar atmosphere. In addition to the "layering" of the solar atmosphere into a chromosphere, transition region, and corona, the relative brightnesses and temperatures of these layers vary with the particular feature of interest—one can, in effect, regard each feature as having its own atmosphere. Each of these has its own characteristics in temperature, density, magnetic field, and variation with the solar cycle.

Solar Active Regions

Active regions are nests of activity that are commonly, but not always associated with sunspots. The active regions are associated with localized regions of strong magnetic field, and the energy in this magnetic field is responsible for the enhanced activity that shows itself as bright emission across the electromagnetic spectrum. As a result of excess magnetic energy, active regions tend to be the homes of strong plasma heating, fast plasma motions, and highly energetic phenomena such as solar flares. The structure of the magnetic field and the enhanced plasma heating associated with it determine the appearance of active regions as concentrations of bright loop-like structures where the plasma is typically denser and hotter, and so more readily observed. Like sunspots, active regions typically form between latitudes 40N and 40S, known as the *activity belt*, and vary with the solar cycle.

The Quiet Sun

In contrast to the large-emission, strong magnetic field and dramatic energetic phenomena of active regions, the regions of the Sun outside of active regions are known as the quiet Sun. It should be pointed out that nowhere is the Sun quiet—everywhere you look the Sun is highly structured, dynamic, and energetic. The quiet solar network encompasses vast regions of small-scale magnetic field that form a web-like pattern observable in the red hydrogen line (Hα) and the ultraviolet line of calcium (Ca II – calcium atoms with one electron removed). The network is defined by the fluid motions of the large-scale photospheric supergranules. Energy released from the quiet Sun magnetic field results in a range of dynamic processes and explosive events including X-ray bright points, nanoflares, and plasma jets.

Spicules

The chromosphere is permeated by a series of short-lived jets of plasma known as spicules. These narrow jets, only 500 km across, burst from the chromosphere at speeds of up to 80,000 kilometers per hour to reach heights of about 7,000 km. Occurring roughly every five minutes and lasting only around ten minutes, spicules are a crucial component of what drives the solar wind (Chapter 8). Recent results suggest that spicules are formed by sound waves generated by the global oscillations of the Sun (Chapter 4) that leak into the chromosphere to create shock waves. The shock waves then drive the spicules and eject mass into the solar wind.

Coronal Holes

The poles of the Sun tend to be large regions devoid of emission at coronal temperatures. Such voids are evident in eclipse observations and even more startling in the X-ray images taken by telescopes on board the Skylab and Yohkoh missions (Figure 7.4). Such voids were named *Koronale Löcher* (coronal holes) by Swiss solar scientist Max Waldmeier (1912–2000) in the 1950s. Coronal holes are now known to be regions of the Sun where the

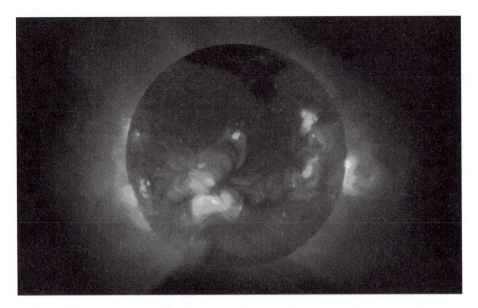

Figure 7.4 X-ray corona showing large coronal hole at the north pole. Source: Solar X-ray images are from the Yohkoh mission of ISAS, Japan. The X-ray telescope was prepared by the Lockheed-Martin Solar and Astrophysics Laboratory, the National Astronomical Observatory of Japan, and the University of Tokyo, with the support of NASA and ISAS.

magnetic field lines extend into interplanetary space. This allows the plasma to readily escape, leading to low density and, consequently, low emission. The temperatures in coronal holes are around 1 million Kelvin, comparable to the bulk of the corona. Their dark appearance is therefore mostly a density effect. The chromospheric portion of coronal holes is similar in nature to that of the quiet Sun, with small bipolar magnetic fields, small bright points, and a network of chromospheric emission. Coronal holes are the source of the fast solar wind (see next chapter). During solar minimum, coronal holes are generally confined to the polar regions of the Sun. However, as activity increases towards solar maximum, the magnetic field of the active regions interacts with the polar field to create extensions to the polar coronal holes that can stretch all the way to the equator. One such example is shown in Figure 7.4.

Helmet Streamers

Eclipse photographs of the solar corona show a wide array of rays, or streamers, emanating from the Sun in all directions and essentially along radial lines, out to several million kilometers. Some of these streamers are associated with low coronal structures such as active regions and solar prominences, which result in a characteristic helmet shape—thus these streamers are called *helmet streamers*. Helmet streamers are regions of enhanced coronal density and so appear brighter than the rest of the corona in eclipse pictures. The mass that makes up this density is supported against gravity by the Sun's magnetic field. The helmet streamers extend to coronal heights where the solar wind can essentially stretch out the magnetic field lines to form the characteristic "spike" at the top of the helmet. Coronal plasma is free to flow along this spike or streamer into interplanetary space, contributing to the equatorial solar wind. During solar minimum, the helmet streamers form a continuous belt around the Sun, known unsurprisingly, as the streamer belt. The extensions of this belt into interplanetary space define a heliospheric current sheet, which we will discuss in Chapters 8 and 10.

Magnetic Loops

Clearly, the most prominent features in the low solar atmosphere are the magnetic loops that cover the Sun, which are most readily observed in solar active regions. These loops are the result of heated plasma lying along magnetic field lines that connect regions of opposite polarity on the Sun, across what is called a magnetic neutral line (see Chapter 3). Because of the strength of the magnetic field, relative to that of the plasma, in the solar corona, the plasma is confined to these elongated loop structures and provides a picture of the otherwise unobservable magnetic field in the corona. These loops also provide important clues to how the atmosphere of the Sun is heated and how the energy is distributed.

Simple semicircular structures, complex S-shaped and cusped (pointy) loops, and loops arrayed along an arcade make up the zoo of structures associated with heated plasma and magnetic neutral lines. All of these structures tell us something about the physical interaction between the magnetic field and solar plasma and lead to a better understanding of coronal heating and energy release processes that can have significant consequences for us here on Earth.

Sigmoids

A common, and important, feature often seen in solar active regions is the *sigmoid*, named after its characteristic S-shaped structure. The presence of such structures was first noted in X-ray pictures of the Sun, but sigmoids have since been identified at other coronal wavelengths, particularly EUV. Sigmoids have been associated with the production of solar flares and coronal mass ejections (Chapter 9). They are regions of excess magnetic energy driven by oppositely directed motions on either side of the magnetic neutral line (so-called shearing motions).

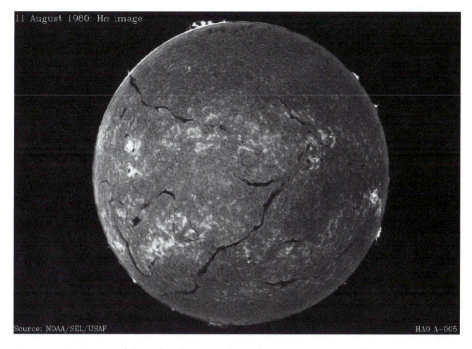

11 August 1980: Hα image

Source: NOAA/SEL/USAF

HAO A-005

Figure 7.5 Hα image of the solar chromosphere showing prominent filaments across the solar disk. Courtesy of NOAA.

material clearly marks it as chromospheric, yet it lies high above the normal chromospheric layer. The key again is the magnetic field.

Prominences form along the magnetic neutral line separating two distinct magnetic polarities (Chapter 3). The dense prominence material is supported in the corona against the force of gravity by the presence of elongated magnetic field where the field lines are either twisted around, like a slinky, with the prominence plasma settling in the bottom sections of the twists, or have marked dips where the plasma can settle and accumulate.

While prominences are most noticeable off the limb of the Sun where they are seen as bright Hα objects in emission, they can also be seen on the disk of the Sun. Because of the bright emission of the solar chromosphere, which overwhelms the emission from the prominence, the prominence material absorbs some of the background Hα emission (see Chapter 2) and shows up as an elongated dark feature. On the disk, a prominence is known as a filament (Figure 7.5). While the prominence and filament look very different, they are actually one and the same thing.

Prominences and filaments can form in and around active regions, where they are known as *active region prominences*, or in the quiet Sun, particularly near the polar crown, where they are known as *quiescent prominences*. Typically, solar prominences will last for several days or weeks, surviving for several solar rotations. Frequently, however, prominences undergo a dramatic change where they expand and erupt out into space, leaving behind a solar flare and generating a coronal mass ejection that can hurtle away from the Sun at several thousand kilometers per second. The processes and conditions that lead a prominence to erupt is still unknown and is one of the most active areas of solar physics research. We will return to this puzzle in Chapter 9.

CORONAL HEATING

The laboratory identification in the 1930s that a mysterious line detected in the solar corona was actually a highly ionized form of iron (see sidebar) pointed to the rather remarkable conclusion that the corona was hot. The reason that this was surprising was that it had long been known that the surface was around 5800 K, and so it was hard to imagine a source of heat that could lead to a temperature of 1 million degrees in the diffuse corona. Imagine trying to get hot water to the far end of the bathtub while keeping the water at the faucet end cold. This is the coronal heating problem.

Maintaining a hot million degree corona, which is in direct contact with a 10,000 degree chromosphere and a 6,000 degree photosphere, is not easy. Not only do you have to heat the plasma there, but you also have to keep heating it to balance the losses by radiation and conduction (see Table 7.2). Your soup gets cold if you leave it too long as it conducts heat into the bowl and radiates heat out of the top of the bowl.

Table 7.2 shows how much energy per square meter the different regions of the solar atmosphere lose every second. To maintain the observed temperature of the different regions, heat must be supplied to balance these losses. In the corona, the conductive flux represents the amount of energy that flows from the hotter parts of the atmosphere to the cooler parts (i.e., from corona to chromosphere), the radiative flux represents the amount of energy radiated away, and the solar wind flux represents how much energy

Table 7.2. **Energy losses exhibited by the solar atmosphere**

	SOLAR ACTIVE REGION	QUIET SUN	CORONAL HOLE
Corona			
Conductive Flux	100 − 10,000	200	60
Radiative Flux	500	100	10
Solar Wind Flux	<100	<50	700
Total Energy Loss	∼10,000	∼300	∼800
Chromosphere			
Radiative Flux	∼20,000	∼4000	∼4000
Solar Wind			
Mass Loss	$<4 \times 10^{-10}$	$<2 \times 10^{-10}$	$\sim 2 \times 10^{-9}$

Note: Energy flux is in units of Watts per square meter and mass loss is in units of kilograms per square meter per second is carried off by the solar wind. In the chromosphere, the dominant loss is due to radiation. The table readily shows that, in spite of the much higher temperature of the corona, the amount of energy required to maintain the chromosphere at ~10,000 K is 2–10 times higher than required to heat the corona. This is primarily due to the much larger amount of material per unit volume in the chromosphere compared to the corona. The reason there is no "chromospheric heating problem" is that there is ample energy provided by convective motions and the magnetic field to adequately overcome the large chromospheric energy losses. Dissipating enough energy in the much less dense corona is extremely difficult and has provided a challenge since the corona was first discovered to be hot.

Discovery of Coronium

During observations of the solar eclipse of August 7, 1869, American astronomers Charles Young (1834–1908) and William Harkness (1837–1903) independently detected a single, very strong, green emission line in the spectrum of the solar corona. This line had not been detected from any known element in the laboratory, and so it was thought that a new element, named *coronium*, had been discovered. This was followed in subsequent eclipse observations by the detection of several more spectral lines that were unique to the solar corona, and not associated with known laboratory emissions. These lines remained unidentified until the late 1930s when spectral studies by German physicist Walter Grotrian (1890–1954) and Swedish physicist Bengt Edlén (1906–1993) found that they were due to highly ionized iron, nickel, and calcium—the green line at 530.3 nm, in particular, is due to iron ions that have been stripped of thirteen of their twenty-six electrons. Similar results applied to the other coronal lines detected. The highly ionized states of these elements can only be achieved at extremely high temperatures, in excess of 1 million Kelvin. There was now no escaping the fact that the solar corona was extremely hot. How could this be when the temperature had been declining from the core to the surface and the corona should therefore be even cooler? The coronal heating problem was born.

The first step to tackling the coronal heating problem is to determine how much heat is needed and how this heat is distributed between the particles in the solar atmosphere. To understand this distribution of temperature and density in the solar atmosphere, solar physicists employ a number of simplifying assumptions. One of the most useful, although in many cases clearly wrong, is that of assuming a hydrostatic atmosphere. The *hydro* here stands for fluid, which in the case of the solar atmosphere is simply the plasma that makes it up, and the *static* stands for not moving. In other words, a hydrostatic atmosphere is one where the atmosphere has settled, under gravity, into a stable, unchanging distribution of density and temperature. We have emphasized in previous chapters how the solar atmosphere is constantly changing, so this assumption seems to be unrealistic. However, in many cases the times over which changes occur are sufficiently long that the hydrostatic assumption can be used to obtain a snapshot of solar conditions over times shorter than the timescale for the conditions to change. In the solar atmosphere, this time varies significantly but can typically be around thirty minutes to one hour.

In a hydrostatic atmosphere, the pressure of the plasma is balanced by the gravitational forces acting on it, resulting in a distribution of density that decreases with height. On the Earth, this can be seen, and felt, by the increased difficulty in breathing as one climbs a high mountain. Mountaineers climbing the highest peaks typically carry oxygen tanks to compensate for the thinning of the air. As a result of gravity, gas at low heights in the atmosphere is compressed more by the gas above it. As one goes higher, there is less gas above and so less compression, and the density gets lower. The density falls off uniformly with height. How quickly the density falls is called the *density scale height*, which measures the distance over which the density falls by 1/e, with the number e = 2.71828 being the base of the natural logarithm. On the Earth the density scale height is about 4 km, while in the solar corona it is around 50,000 km due to the higher temperatures and the fact that the corona is mostly hydrogen, which is lighter than the nitrogen/oxygen mixture of air on Earth.

The balancing of gravity forces with the pressure forces creates a very simple distribution of density within the atmosphere for a given temperature, which can then be compared to observations to determine the state of the solar corona. The key to this is the balance between the different mechanisms involved. From our description of a hydrostatic atmosphere we see that the pressure distribution can be determined by balancing gravity at every height and assuming there are no extraneous motions. However, there is more to this than one might think. Even if the atmosphere is able to "ignore gravity," the high pressures required means that the atmosphere is hot, and the distribution of temperatures means that some parts of the atmosphere are hotter than others. Now, hot plasma radiates and loses heat to space (think of a hot ember in a campfire glowing and giving off light) so it naturally wants to cool. Moreover, the hotter parts of the atmosphere will want to share their energy with the cooler parts (think of cooling down your bath by running cold water into the tub). This latter effect is known as conduction. Both of these effects mean that the atmosphere is losing energy and that the conditions are changing. The atmosphere can still be kept in balance if there is an additional source of heating to counteract the cooling of the plasma to radiation and conduction. This is reflected in the following *energy equation* for a hydrostatic atmosphere:

$$E_h = E_r + E_c$$

where E_h is how much heating is being injected per second into the part of the atmosphere of interest, E_r is how much is being radiated away, and E_c is how much is being conducted away to other parts of the atmosphere. The equal sign here means that the losses (E_r and E_c) are being exactly balanced by the gains (E_h) in energy. When combined with the hydrostatic balance, this completely defines the physics of this simplified atmosphere. The important part here is that the energy losses to radiation and conduction are well understood, so by observing the temperature and density distribution in the solar atmosphere, we can work out how much heating is required to keep the atmosphere in balance, and so know how much energy needs to be supplied to keep the corona hot.

This helps tackle the coronal heating problem, but unfortunately there are so many possibilities that we can never be sure that we are correct. Add the fact that the plasma in the atmosphere also moves energy around by flowing between one place and another, and the problem becomes even harder.

While there is still much discussion on the specific mechanism of heating the solar corona, it is clear that the energy provided by the magnetic field that permeates the solar atmosphere is crucial. The magnetic field not only acts as a conduit for plasma in the corona, confining it to unique structures called loops, but also acts as a source of energy that gets released when the field interacts with the solar plasma. It is this interaction that remains a mystery. There are many mechanisms proposed for heating the corona, but the bulk of them fall into one of two categories: heating by wave action, where the field wiggles back and forth with some of the wave energy being transferred to the plasma as heat; or heating by currents, where the motions of the solar surface twist and wind up the magnetic field, creating electric currents (see Chapter 3) which then interact with the ambient plasma dissipating their energy. Observations have been critical in driving the coronal heating debate, but as yet a definitive theory remains elusive.

The solar atmosphere is a complex system involving the interaction of magnetic field and plasma. Observations have shown us the complexity and dynamism of the chromosphere, transition region, and corona, connecting it to the dynamics and radiative output of the solar surface. The atmosphere does not stop in the immediate vicinity of the Sun, but spreads throughout the whole solar system and beyond, only terminating where the pressure of the solar wind matches that of the surrounding stars. The "bubble" that separates the Sun from the stars is known as the heliosphere, a region some 28 billion kilometers across with the Sun at the center. The two *Voyager* spacecraft that left the Earth over thirty years ago in 1977, traveling at 38,000 miles per hour, have yet to reach the outer boundary of the heliosphere, known as the *heliopause*. In fact, the spacecraft are not expected to cross the heliopause for another twenty years.

All the planets and bodies of the solar system are connected to the solar atmosphere through their interaction with the solar wind, the topic of the next chapter, and it is this interaction that makes an understanding of the solar atmosphere and its variations so important. Discovering how the atmosphere is heated, how the dynamical variability is mediated through the various atmospheric layers, and how the interplay of magnetic field and plasma lead to the sudden eruptions of solar storms is one of the most intriguing, and arguably most important for us here on Earth, mysteries of the Sun. Understanding the solar atmosphere will not only better define the interaction between the Sun and the Earth, but will also provide insight into the atmospheres of other stars and their interactions with their planetary systems. The latter may prove critical in our search for life on other worlds.

RECOMMENDED READING

Golub, Leon, and Jay M. Pasachoff. *The Solar Corona*. Cambridge: Cambridge University Press, 1997.

Tandberg-Hanssen, Einar. *The Nature of Solar Prominences*. Dordrecht: Kluwer Academic Publishers, 1995.

"Total Eclipse Solar Eclipses and the Mysteries of the Corona." Exploratorium and NASA's Sun-Earth Connection Education Forum. DVD. NASA: Greenbelt, MD. 2003.

WEB SITES

Mauna Loa Solar Observatory: http://mlso.hao.ucar.edu.

SOHO/LASCO: http://lasco-www.nrl.navy.mil.

TRACE: http://trace.lmsal.com.

8

Blowing in the Wind

The hot solar atmosphere discussed in the previous chapter has some interesting consequences that extend well beyond the immediate environment of the Sun. Like most hot gases, the solar atmosphere is continuously moving in all directions, confined to the Sun by gravity and the confining presence of the Sun's magnetic field (Chapter 3). However, a number of interesting phenomena suggested that something else was going on. As early as 1619, the now-famous German mathematician Johannes Kepler (1571–1630) suggested that the tails of comets were formed by the pressure of sunlight pushing on the particles of the comet. This, naturally, explained the propensity for comet dust tails to point directly away from the Sun. In the seventeenth century, light was believed to consist of particles (or *corpuscles*) and so the presence of cometary tails suggested the presence of a stream of particles emanating from the Sun and traveling through interplanetary space. However, in the early nineteenth century, observations showed comets displaying two distinct tails, one continuously pointing away from the Sun and the other changing its direction as the comet moved around in its orbit. This argued against the simple idea that the pressure of light particles from the Sun was responsible, since this pressure would be constant and only create a single tail pointing directly away from the Sun.

The first person to predict that the Sun must indeed be the source of a constant stream of fast particles was Norwegian scientist Kristian Birkeland (1867–1917) in 1916. Birkeland was trying to explain the presence of the terrestrial aurora (or northern lights; there is also an equivalent southern lights seen at large southerly latitudes). The aurora had earlier been found to come and go with the presence of sunspots, and to vary with the solar cycle, suggesting a direct connection between these terrestrial phenomena and events on the Sun. (More about this connection will be discussed in Chapter 10.) Birkeland proposed that charged particles (electrons, protons, and ions) produced by the Sun traveled through the solar system, and that some of these particles got "caught up" by the magnetic field of the Earth, which funneled them towards the magnetic north and south poles (see Chapter 3) where they interacted with the charged particles of the Earth's atmosphere to produce the fascinating light display of the aurora. Birkeland was able to demonstrate his theory by conducting experiments in the laboratory that showed that electrons generated by an electrical source were channeled towards the poles of a magnetic sphere where they produced an array of glowing shapes in the phosphorescent paint coating the sphere.

In addition to the experiments and hypotheses of Birkeland, observations of comets throughout the nineteenth and early twentieth centuries confirmed the presence of the two tails: one tail was composed of dust and was directed directly away from the Sun, while a second tail was composed of ions and curved around in the direction of the comet's orbit. The ion tails were also found to be blue in color and to exhibit a range of behavior not consistent with being pushed by sunlight. In 1943, German scientist Cuno Hoffmeister (1892–1968), and subsequently Ludwig Biermann (1907–1986) in 1954, proposed that the Sun produced a steady stream of particles, which had sufficient momentum to push the ions of the comet away from the nucleus. The curved path of the ion tails, back along the orbital path of the comet, meant that the ions were being kicked off the comet nucleus with low energies, suggesting that the solar particle stream was moving relatively slowly, a few hundred kilometers per second.

While the observational evidence pointed to a clear presence of a continuous stream of particles emanating from the Sun, there was no physical rationale for why such a stream should exist. This changed in 1958 when American astrophysicist Eugene Parker decided to have a closer look at the dynamic structure of the solar corona.

PARKER'S MODEL

The basic tenet behind Parker's explanation of how particles can stream from the Sun continuously is based on the fact that a static, that is, non-moving, isothermal solar atmosphere cannot balance the gravitational attraction of the Sun and simultaneously be in balance with the surrounding interstellar medium. The solar corona, like the Earth's atmosphere, is constantly fighting gravity to remain in place. The net effect is that as you move further away from the Sun, the density falls, which, in turn, means that the plasma pressure decreases as you go to higher heights. For the atmosphere to be static and unmoving, the forces resulting from this change in pressure with height have to balance the pull of gravity everywhere in the atmosphere. Parker found that the result of such a solution was that the pressure of the solar atmosphere at very large distances from the Sun was too high; formally even at an infinite distance from the Sun there was a substantial pressure associated with the solar atmosphere. This was unsatisfactory and led Parker to propose that the solar atmosphere could not be static, but must be in a state of continual expansion. The additional forces associated with this flow meant that the pressure forces did not have to be so large, and the pressure could fall off more rapidly, eventually reaching a balance with the surrounding interstellar medium. Parker named this continuous outflow the *solar wind.*

Parker Solar Wind Model

For a static solar atmosphere, the balance of gravity with the forces due to the change in pressure with height can be expressed as:

Force of gravity = pressure gradient forces   or
$-GM\rho/r^2 = dp/dr$

where the gravitational force is dependent on the mass of the Sun, M, the density of the atmosphere, ρ, and falls off with the square of the distance, $1/r^2$. The parameter G is a constant number known as the gravitational constant, and the negative sign reflects the fact that gravity is acting downwards. The term on the right hand side of the equation just reflects that the change of pressure with height is given by the need to balance the gravitational force on the left hand side.

For a solar atmosphere at temperature, T, the density, ρ, is related to the pressure, p, via the ideal gas law:

$\rho = mp/(2kT)$

where m is the mass of protons in the atmosphere and k is another constant, called the *Boltzmann constant.* Using the ideal gas law, the force balance equation above can be solved to show that the pressure falls off with distance like an exponential function, which at large distances from the Sun looks like:

$p = p_0 e^{-\lambda/Rs}$

where R_s is the solar radius and λ is a characteristic length scale associated with the temperature, T, and p_0 is the pressure at the base of the corona. Thus, for the temperature and density at the base of the corona, we have p_0 and $\lambda = 47,000$ km. Put all this together and the pressure at a large distance from the Sun turns out to be 10^{-13} N m^{-2}, which is larger than observations at 1 AU and too large to be in equilibrium with the interstellar medium.

Parker's solution was to add a flow speed variation to the balance equation such that *Force of gravity = pressure gradient forces + velocity gradient forces* or

$-GM\rho/r^2 = dp/dr + \rho v(dv/dr)$

allowing the pressure to fall to arbitrarily small values at large distances.

The solar wind, then, is a stream of charged particles that continuously flows radially outward from the Sun in all directions to fill interplanetary space. The solar wind is responsible for several solar system phenomena including cometary ion tails, the production of magnetic storms on the Earth, Jupiter, and Saturn, and large-scale structuring of the Sun's magnetic field (see Chapter 3). The solar wind is not uniform, and recent observations are showing that it is a complex regime of plasma and magnetic field with varying speeds, a puzzling acceleration, and complex interactions.

Parker Spiral

The coronal magnetic field and the plasma properties of the solar wind are intimately related. Charged particles, such as the ions, protons, and electrons that make up the solar wind have great difficulty crossing magnetic field lines, but easily flow along them (Figure 8.1). Thus, in regions where the field is strong and horizontal, relative to the radial solar wind flow direction, the wind is hindered. On the other hand, where the field is radial and extends out into interplanetary space (the open field lines discussed in Chapter 3), the solar wind flows readily, and because of the large conductivity of the solar corona, the solar wind drags the solar magnetic field with it to form the interplanetary magnetic field (IMF) that pervades the whole solar system. Even though the solar wind flows radially outwards from the Sun, the rotation of the Sun introduces an interesting pattern in the IMF. For a typical slow solar wind flow speed of about 400 km/s, the solar wind takes about 4.3 days to reach the orbit of the Earth, 150 million km away. During this time, the Sun rotates about 55° westwards since the Sun, at its equator, rotates on its axis every twenty-seven days. This combination of solar rotation and solar wind outflow makes the interplanetary magnetic field form a well-defined mathematical shape known as an Archimedean spiral, or in the solar

Figure 8.1 Representation of the Parker spiral. Courtesy of NASA.

context a Parker spiral, since it was first pointed out by Eugene Parker in his 1958 solar wind paper. The presence of the Parker spiral and the fact that charged particles like to remain tightly confined to magnetic field lines means that only locations close to the western limb of the Sun (the right side of the Sun as viewed from the Earth) are physically connected to the Earth. The exact western longitude varies with the variations in the solar wind speed. This has important consequences for space weather (see Chapter 10).

OBSERVATIONS

The observational confirmation of the solar wind came from spaceborne particle detectors. Ion trap detectors on an early Russian rocket flight, *Lunik 2*, showed tantalizing evidence for an increase in the electrical charge detected whenever the spacecraft pointed towards the Sun. Similar observations were found by NASA's *Explorer 10* mission in 1961. However, it wasn't until space missions that could escape the messy particle environment of the Earth's magnetosphere that the real evidence for the existence of a solar wind became available. In 1962, NASA launched the *Mariner 2* spacecraft on a mission to the planet Venus. *Mariner 2* not only detected a continuous flow of charged particles originating at the Sun, but also found that the flow was comprised of a complex mixture of fast and slow streams. The detected proton density varied from 1 million particles per cubic meter to over 50 million particles per cubic meter, and the speed ranged from about 300 km/s to as high as 800 km/s, with the lower densities being associated with the highest speeds. This fast/slow stream pattern repeated every twenty-seven days, suggesting a tie-in to solar rotation and therefore implying that they were solar in origin. The sources of the fast streams within the solar wind were found later to have their origins in coronal holes, regions on the Sun where the magnetic field is open to interplanetary space (see Chapters 3 and 7).

As observations improved with the development of better instrumentation, the solar wind was found to have a whole array of interesting physical properties. Measurements established that the bulk of the solar wind consisted of electrons and protons, with helium only accounting for 5% of the material in the fast streams, and heavier ions only being detected in trace amounts. The various species making up the solar wind were found to have widely different temperatures. One of the most important discoveries was that there were two distinctly different solar winds, best characterized by their velocity: *the fast solar wind* and *the slow solar wind*. The properties of these two distinct winds are summarized in Table 8.1. In addition to the large difference in flow speed, the properties of the slow solar wind are very different from those of the fast solar wind.

The launch of the *Ulysses* spacecraft in 1990 provided the first set of in situ measurements out of the ecliptic plane (the plane containing the Sun

Table 8.1. **Properties of the solar wind**

Wind Property	Fast Solar Wind	Slow Solar Wind
Solar origin: solar minimum	Coronal holes	Streamer belt ±35° latitude
Solar origin: solar maximum	Polar and low latitude coronal holes	All latitudes
Nature of wind	Steady and uniform	Highly variable
Average speed	700 km/s	350 km/s
Temperature of protons	280,000 K	55,000 K
Temperature of electrons	130,000 K	190,000 K
Temperature of helium ions	730,000 K	170,000 K
Density of protons (at 1 AU)	∼3 per cubic centimeter	5–10 per cubic centimeter

and the orbit of the Earth). *Ulysses* is the only spacecraft to orbit the Sun in an out-of-the-ecliptic orbit, traveling from the equator to above the solar poles with an orbital period of 6.3 years. At its closest point *Ulysses* is about 30% further away from the Sun than the Earth, and at its furthest point it is about 5.3 times further away, close to the orbit of Jupiter. The unique orbit of *Ulysses* allowed it to detect the variation of solar wind properties with latitude as it passed over both the southern and northern poles of the Sun. The pole-to-pole measurements of *Ulysses* have now spanned almost two decades and have provided a wealth of scientific information on the properties of the solar wind and their variation with time.

The two-fold nature of the wind emanating from the Sun is exemplified by Figure 8.2, which shows the combined results of the *Ulysses* fast latitude scans over the period 1992 to 2003, spanning a complete solar cycle. Figure 8.2 shows several interesting phenomena. First, either of the two upper plots indicates a strong variation of the solar wind speed with latitude. This is shown best on the upper left panel. The solar wind speed at low latitudes near the equator is much slower than the speed measured away from the equator and towards the poles. The fast and slow solar winds therefore emanate from high and low latitudes, respectively. Comparing the declining phase of the solar cycle (upper left panel) with solar maximum (upper right panel), it can be seen that near solar maximum the solar wind speed is highly variable and the fast and slow streams are much more intermingled. It is, therefore, much more difficult to decouple the fast and slow winds that are so evident during relatively inactive periods of the solar cycle.

One of the more notable discoveries is that the structure of the solar wind is very different at solar maximum than it is at solar minimum. At solar maximum, there is a clear north-south asymmetry (see Figure 8.2, above). In the southern hemisphere the fast wind is almost entirely absent, while the variable slow solar wind is present at all latitudes. In the northern hemisphere, the fast wind is seen at high latitudes (>70°).

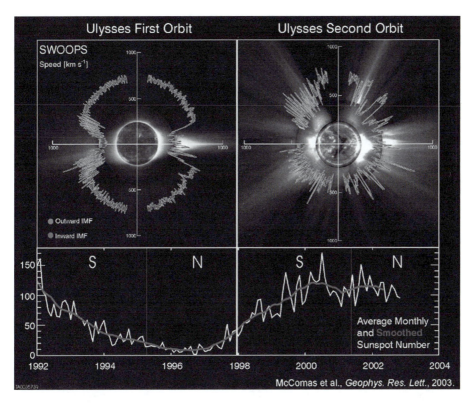

Figure 8.2 Observations of the solar wind from the first (near solar minimum) and second (near solar maximum) orbits of the *Ulysses* mission. The polar plots at the top of the figure show the solar wind speed as a function of latitude. Courtesy of Dave McComas.

The varying presence of the fast solar wind is directly related to the presence of coronal holes. During solar minimum, the coronal holes are found predominantly at the poles of the Sun, and the fast solar wind is well-structured and concentrated at high latitudes. As the Sun moves towards solar maximum, the interaction between the increasing number of active regions and the polar fields results in the generation of small coronal holes at low latitudes or large equatorward extensions of the polar coronal holes (see Figure 7.4) all of which result in the mixing of fast solar wind streams into the background slow solar wind associated with the activity belt streamers.

A hint towards the source of this difference can be seen in the superimposed coronal images in Figure 8.2, above. On the upper left the solar corona is extremely simple, with a predominantly dipolar structure (see Chapter 3)—the slow wind is associated with large helmet streamers that separate closed and open field, while the fast solar wind is associated with the polar open fields (coronal holes). At solar maximum, the coronal structure is significantly more complicated, with large streamers evident at both low and high latitudes. The coronal hole structure extends equatorward, resulting in the intermingled fast and slow streams.

Solar Wind Mass Loss

The continuous outflow of protons, electrons, and ions that make up the solar wind constitutes a significant, and relentless, removal of material from the Sun into interplanetary space. Given the average density of around 5 million particles per cubic meter of the solar wind, and the average speed of around 450 km/s, one can calculate that approximately 1 billion kilograms of solar plasma is carried away from the Sun by the solar wind every second (solar wind mass loss is 10^9 kg/s). While this sounds like a lot, the sheer mass of the Sun means that at this rate of mass loss it would take around 10 trillion years (10^{13} yrs) for the Sun to completely blow itself away into space. The mass loss of the Sun is dominated by the nuclear fusion process in the core that converts hydrogen to helium, with a mass loss rate of five times that carried off by the solar wind. It has been suggested for star formation theory that in the distant past when the Sun was a young star, the solar wind was as much as 1,000 times more massive.

The solar wind affects planets in two major ways, depending upon whether the planet has its own magnetic field or not. The atmospheres of planets without an appreciable magnetic field (Venus and Mars) are exposed to the full force of the solar wind, which can result in the "stripping" of the atmosphere of the planet. The strength of this effect, relative to that caused by the heating due to solar optical and UV radiation, depends crucially on the distance of the planet from the Sun, the mass of the atmosphere, and the mass of the planet. In essence, the solar wind "blows away" some of the planetary atmosphere. Modern spacecraft have found some evidence for a comet-like tail extending from Venus's nightside. Importantly, the fact that the solar wind may have been 1,000 times more massive in the early history of the solar system is thought to be responsible for the relatively small atmosphere on Mars and the resulting drying up of that planet's surface.

For the planets with a magnetic field, the effects of the solar wind are more spectacular if less damaging to the planet itself. The magnetic field of a planet effectively shelters the planet from the solar wind by channeling the charged particles away towards the poles. (Charged particles do not travel easily across magnetic field lines, but can readily travel along them; see Chapter 3). This magnetic blanket has protected the Earth's atmosphere from being eroded by the solar wind streaming past it. However, the interaction of the planetary magnetic field with the interplanetary magnetic field associated with the solar wind (see the Parker spiral sidebar) produces a number of interesting effects that are becoming increasingly more relevant as we utilize space in our everyday lives (weather satellites, telecommunications, etc.). The Earth (and Jupiter, Saturn, Uranus, and Neptune) are all surrounded by a bubble of magnetic field, called the *magnetosphere*, due to electromagnetic forces within their interiors (Chapter 3). This bubble is affected by the solar wind in two ways. First, the solar wind pressure pushes on and distorts the bubble, compressing the magnetosphere on the sunward side and forming a long extended tail in the anti-sunward direction. Changes in the solar wind cause the magnetosphere to fluctuate significantly, changing local conditions in space—a large compression can push the magnetosphere so low that satellites can find themselves exposed directly to the solar wind, with damaging consequences. Second, the charged particles of the solar wind that are channeled to the poles by the planetary magnetic field interact with the upper atmosphere of the Earth to create an array of electromagnetic phenomena such as aurora and geomagnetic storms. Aurorae have been observed to occur at the poles of both Jupiter and Saturn. The interaction of the solar wind with the Earth's atmosphere and its effect on society will be discussed in much greater detail in Chapter 10.

SOURCES OF THE SOLAR WIND

The development of our understanding of the interaction of the solar wind with the Earth goes back even before the solar wind itself was discovered. As early as 1927, a twenty-seven-day periodicity was detected in the occurrence of geomagnetic storms. Geomagnetic storms had been known to be associated with large solar flares since the mid-1800s (see Chapter 9), but the periodic occurrence of storms indicated the presence of an additional solar source of energetic particles. The twenty-seven-day period pointed to a solar origin as it closely marked the solar equatorial rotation rate, as seen from the Earth. The recurrent storms were not as strong as the flare-associated ones, but were significant, nonetheless. German scientist Julius Bartels (1899–1964) found recurrent sequences in the storms, which he clearly illustrated by *Bartels diagrams* starting in 1932. Bartels associated the recurrent storms with hypothetical regions that he called *M regions*. Despite all of his efforts, he was unable to find any correlation of the recurrent storms with any visible feature on the Sun. Statistical studies showed that the M regions could not be associated with sunspots or streamers. An important piece of evidence regarding the nature of the M regions came from a study of eight M storms that occurred in 1950 and 1951. Astronomer M. J. Smyth of the Royal Observatory in Edinburgh found that the onset of each of these storms followed the passage of unusually faint regions of the corona across the central portion of the solar disk. These regions were later identified, from the Skylab mission in the early 1970s, with what we know today as coronal holes. What was clear was that geomagnetic disturbances, large and transient or small and recurring, required the production of particles (the "corpuscles") at the Sun and their subsequent transport to the Earth. From this relatively simple fact, the concept of a "corpuscular wind" was developed.

The identification of the source of the solar wind is only part of the puzzle. How the solar wind gets accelerated and where the energy for the acceleration comes from is still an active area of research, with some important clues coming only in the last decade. Observations from the SOHO spacecraft have shown that small-scale structures at the base of coronal holes are

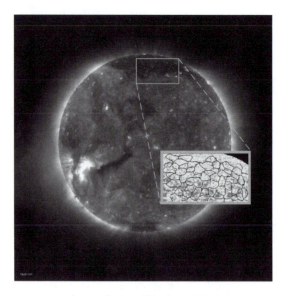

Figure 8.3 Velocity structure at base of coronal hole. Full Sun image in the EUV from the Extreme Ultraviolet Imaging Telescope (EIT) on SOHO. Velocity structure map from SUMER spectrograph on SOHO. Courtesy of the SOHO. Credit: ESA/NASA.

the source of high-speed outflows, and that the fastest flows are concentrated around the boundaries of the magnetic network (Chapter 3). The magnetic field in these regions forms open tubes that eject the solar wind plasma out and away from the Sun.

Figure 8.3 shows the results reported by Don Hassler of Southwest Research Institute in Boulder, Colorado, using data from the SOHO SUMER instrument that can measure the Doppler line shifts caused by the motion of the plasma. The Doppler velocity data show a correlation between the fastest upflows and the vertices of the magnetic structures observed on the solar surface. Velocities measured are around 10 km/s, which must be continually accelerated as they pass through the corona to reach fast wind velocities of 750 km/s.

In addition to the Doppler evidence just discussed, solar scientists have also found that spiky jets of material, called spicules, are constantly being ejected from the Sun and may form the basic building blocks of the solar wind. The solar surface is filled with literally hundreds of thousands of spicules, each about 500 km across, 5,000 km tall, and ejecting mass at 22 km/s. The spicules come and go on a five-minute timescale and provide sufficient energy to drive the solar wind. Most of the mass ejected is seen to fall back towards the surface, so it is far from certain that this provides the source of the solar wind, but the spicules carry more than a hundred times the mass required to provide the solar wind outflow, so if only 1% of the mass escapes the Sun's gravity, then the puzzle of the origin of the solar wind may be close to being solved.

There remains the problem of how the solar wind is further accelerated to the large velocities measured in interplanetary space. The slow wind speeds near the solar surface are, in part, a result of the plasma having to fight against solar gravity, which is strongest near the surface. As the plasma gets higher into the solar atmosphere, gravity decreases and the wind can pick up speed, ultimately reaching what is known as a terminal velocity, where the force of the acceleration is matched by the force of gravity in the opposite direction. This terminal velocity is the solar wind speed. The problem for the Sun is two-fold as the fast and slow winds have different terminal velocities and, consequently, different sources of acceleration, with the fast solar wind naturally requiring more energy.

Observations from *Ulysses* showed that the slow solar wind originates in hot 2–3 million degree regions, whereas the fast solar wind originates in the relatively cool one million degree plasma regions of coronal holes. Furthermore, the slow solar wind is found to take a long time to accelerate to its terminal speed of around 400 km/s, with the maximum speed occurring at the ends of the streamers some twenty or so solar radii from the surface (~15 million km). The hot expansion of the solar corona envisaged by Parker can readily explain the slow solar wind speeds and its energy content.

The fast solar wind, on the other hand, accelerates very quickly, reaching 750 km/s at a height of around 5–7 million km. The additional energy required to produce such rapid accelerations to such high velocities means that an additional energy source is required. While there are many ideas, to date no consensus has been reached on the mechanism responsible for accelerating the fast solar wind. Many ideas center on the conversion of wave, or turbulent, energy into kinetic energy of the plasma. The constant buffeting of the solar surface and atmosphere serves to generate a wide range of waves traveling along and across the solar magnetic field. If the properties of the atmosphere and the waves match, then the waves can lose energy to the plasma of the atmosphere causing it to accelerate quickly to large velocities. Theoretical and observational studies are ongoing to determine the origins and driver of the solar wind.

The Day the Solar Wind Disappeared

An interesting and rare phenomenon occurred for three days in May 1999—the solar wind that flows constantly from the Sun virtually stopped blowing. From May 10–12, 1999, instruments on board NASA's ACE (*Advanced Composition Explorer*) and *Wind* spacecraft, designed to continually monitor the solar wind as it flows from the Sun to the Earth, detected a major decrease in the solar wind density and speed. ACE is situated right in the solar wind some 1.5 million kilometers upstream from the Earth. The instruments on board these spacecraft can measure the density, speed, composition, and strength of magnetic field in the solar wind. Starting late on May 10, ACE and *Wind* observed a 98% drop in the solar wind density. Because of this decrease, ACE and *Wind* were able to observe electrons traveling directly from the low solar corona. These electrons are normally mixed in with the solar wind and distributed via collisions throughout the heliosphere. The drop in the solar wind density impacted the electron content in the Earth's radiation belt causing it to be severely depleted for several months afterward. The reduction in solar wind pressure caused the Earth's magnetosphere to balloon out as far as the orbit of the Moon, some five-to-six times larger than normal.

No definitive explanation has been given for this rather rare occurrence. The most prominent suggestion is that the lack of solar wind is related to the "flipping" of the solar magnetic field from one magnetic cycle to the next (see Chapter 6). As the field undergoes a transition, there is a sudden burst of slow wind, which results in the low speed and low density observed.

THE EDGE OF THE SOLAR SYSTEM

An important aspect of the solar wind is that it extends the influence of the Sun further into interstellar space than would normally occur in the absence of a wind. The volume of space directly affected by the Sun's wind and magnetic field is known as the heliosphere (Figure 8.4). Our current estimation of the location of the edge of the heliosphere, where the Sun's influence stops and the pristine medium of interstellar space begins, is approximately 120–150 astronomical units, or 18–23 billion kilometers.

The pressure of solar wind prevents the interstellar medium from encroaching on the heliosphere. However, neutral atoms, interstellar dust, and high-energy cosmic rays are all able to penetrate into the heliosphere to interact with the solar wind. The interstellar neutral particles get ionized as they interact with the solar wind and coronal mass ejections, while the

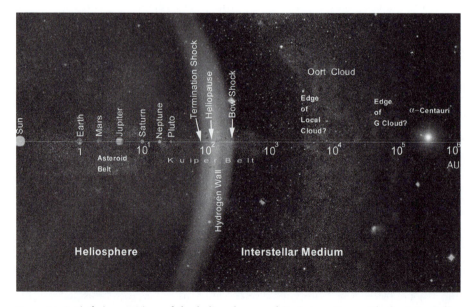

Figure 8.4 Artist's impression of the heliosphere and its interaction with the interstellar medium. Courtesy of ESA/NASA.

cosmic rays are so energetic that they traverse the whole solar system and their effects can even be detected on the surface of the Earth.

Remarkably, we have some direct knowledge of the outer reaches of the heliosphere. In 1977, NASA launched two spacecraft, *Voyager 1* and *Voyager 2*, which became famous initially for their unparalleled views of Jupiter and Saturn. No spacecraft before *Voyager* had traveled so far. The *Voyager* spacecraft also had the distinction that after visiting the outer planets of the solar system their trajectories would take them beyond our solar system to destinations unknown. To commemorate this fact, the *Voyager* spacecraft each carry a twelve-inch gold-plated copper disk, called the *Golden Record*, containing sounds and images portraying the diversity of life and culture on Earth. This includes music by Bach, Beethoven, Louis Armstrong, and Navajo Indians; the sounds of the surf, wind, birds, and whales; and images of human anatomy, planets in our solar system, trees, and various animals. The idea behind this was to communicate information about our world to any extraterrestrial cultures that happened upon the *Voyagers*. Some thirty-one years after launch, traveling at over 38,000 miles per hour, the *Voyager* spacecraft are less than two-thirds the way to the edge of our solar system.

However, twenty-seven years after leaving the environs of Saturn, the science of *Voyager* is coming once again to the fore. Currently, the spacecraft are twice the distance of Pluto. Signals received from the spacecraft indicate that *Voyager 1* has already crossed the termination shock, the location in the heliosphere where the speed of the solar wind changes from being supersonic to subsonic, and that *Voyager 2* is rapidly racing to meet it. Direct measurement of the termination shock will once and for all place a scale on the size of the heliosphere, and truly define where "outer space" begins. The *Voyager* spacecraft are expected to travel into interstellar space sometime after the year 2017.

The solar wind provides the direct and continuous connection between the Sun and the various bodies of the solar system, and beyond into interstellar space. Its influence affects everything from the size of the heliosphere to the stability of planetary atmospheres. At the Earth, the solar wind continually buffets the magnetosphere, generating geomagnetic disturbances that, in turn, have an impact on the Earth itself. The solar wind was discovered almost fifty years ago, and our understanding of its properties, sources, and physical mechanisms has grown exponentially since that time. Yet, there are still many unanswered questions about this most fundamental aspect of the Sun's effect on the Earth and heliosphere. The next fifty years promises to be an exciting period for our developing understanding of the Sun, its wind, and its effect on the Earth.

RECOMMENDED READING

Hassler, Donald M., Ingolf E. Dammasch, Philippe Lemaire, Pål Brekke, Werner Curdt, Helen E. Mason, Jean-Claude Vial, and Klaus Wilhelm. "Solar Wind Outflow and the Chromospheric Magnetic Network." *Science*, 283(1999): 810–13.

Neugebauer, Marcia, and Rudi von Steiger. "The Solar Wind" in J. A. M. Bleeker, Johannes Geiss, and M. Huber, eds. *The Century of Space Science.* Vol. II. Chapter 47. Dordrecht: Kluwer Academic Publishers, 2002.

WEB SITES

The Day the Solar Wind Disappeared: http://science.nasa.gov/newhome/headlines/ast13dec99_1.htm.
Geomagnetic Indices: http://www-app3.gfz-potsdam.de/kp_index/description.html.
Ulysses: http://ulysses.jpl.nasa.gov.
Voyager Spacecraft: http://voyager.jpl.nasa.gov.

9

Solar Storms

The preceding chapters of this volume have discussed the wide range of activity displayed by the Sun and how this activity results from an interaction with the ionized gas of the solar interior and atmosphere (the plasma) and the magnetic field. All of this activity, while important for understanding the Sun as a star and its interaction with the Earth and other planets, is relatively calm, however, compared to the explosive release of energy exhibited by solar storms, the subject of this chapter. Here we use the moniker *solar storms* to represent the array of phenomena associated with the rapid energy release and disruption of the solar corona that results from solar flares and coronal mass ejections (CMEs). Together, the flare/CME phenomena constitute the most dramatic releases of energy in the solar system. These storms are so intense that they can influence events on the surface of the Earth, despite the protection of the atmosphere and magnetosphere. The terrestrial effects of these storms, known as space weather, will be discussed in the next chapter. Here we concentrate on describing the flares and CMEs, their properties, their relationship, and what we know of their cause.

The first solar flare was discovered in 1859, while CMEs were not recognized in their present form until their discovery by the *Skylab* mission in the early 1970s—after the discovery of CMEs, astronomers looked through the historical solar eclipse records and found evidence for CME-like structures propagating through the solar corona as far back as 1860, coincidentally around the same time as the first flare discovery. In discussing the transient energetic phenomena on the Sun, flares have a 115-year advantage over CMEs, and up until relatively recently flares were regarded as the source of enhanced geomagnetic activity at the Earth. Flares were studied in detail from the earliest observations and were primarily regarded as chromospheric phenomena in the earlier part of the twentieth century, but transitioned to predominantly coronal studies with the advent of space-based observations in the late 1950s and later (see sidebar on flare classification in the next section). Since *Skylab*'s discovery of CMEs, flares and CMEs have been studied rigorously, both as separate and integrated phenomena. In the two decades that followed the advent of CMEs, much of the focus was on the question of whether the flare caused the CME or vice versa, the answer to which was hampered by our inability to observe CMEs very close to the Sun. We now know that CMEs are the predominant source of geomagnetic effects whether or not there is a flare at the Sun. The CME, in essence, provides the conduit by which the energetic energy release of a solar storm impacts the Earth. In addition to the impact of the CME, solar flares that are magnetically connected to the Earth along the Parker Spiral (see Chapter 8) can strengthen the geomagnetic impact of the solar storm.

Much of the focus of recent solar research, facilitated by the launch of the SOHO satellite by the European Space Agency and NASA, has been on describing and understanding the impact of coronal mass ejections on the solar atmosphere, solar wind, and interplanetary medium. This is augmented by flare studies that provide direct observational signatures of the energy release at the Sun, with coronal plasma being heated to tens of millions of degrees, and the acceleration of coronal particles to hard X-ray- and gamma-ray-producing energies. Key questions regarding these phenomena include what causes them to occur, how are they related, and can we predict them? The issue is complicated by the fact that both flares and CMEs can, and do, occur independently. The role of the magnetic environment in which the energy release occurs, and the dynamic evolution of this environment, are key factors in determining how the solar corona will respond to a sudden release of energy, governing the production of a flare and/or CME. Frequently, these events are associated with eruptive solar prominences (or filaments), which are magnetic structures supporting cool, dense chromospheric material at coronal heights. Energization of these prominences can lead to a rapid release of the magnetic energy that they contain, resulting in both a flare back at the Sun and a CME that propagates out into interplanetary space. This chapter will describe the flare and CME phenomena and provide insight into their causes, connections, and their impact on the Sun and heliosphere.

Observations of Flares and CMEs

The explosion of solar flare and CME research (pun intended) in the last two to three decades has been fueled by the rapid development in our ability to observe the Sun across the whole electromagnetic spectrum. Flares and CMEs, alike, produce energetic radiation, both particles and photons, which, fortunately for us, cannot penetrate the Earth's protective atmosphere and magnetosphere. The Space Age revolutionized our view of the Sun by enabling us to place

heat to cooler parts of the atmosphere. The duration of each of these phases varies from flare to flare, but typically spans seconds to hours, with the decay phase being significantly longer than the impulsive phase. The particles accelerated in the flare are generally directed towards the dense chromosphere along closed magnetic field lines (Chapter 3), but can also be directed into interplanetary space along open field lines (Chapter 3) where they can contribute to space weather phenomena.

The first-ever solar flare was discovered by English amateur astronomers Richard Carrington and Richard Hodgson (1804-1872) who were, independently, observing sunspots on September 1, 1859, when a sudden brightening of the optical emission occurred (see sidebar). Thus was born the first solar flare. Associated with this remarkable event was a disturbance in the Earth's magnetic field that led some scientists to relate the two. The importance of the chromospheric eruptions, as the early flares were known, for the Earth's space environment came through the study of these events and their apparent association with magnetic storms at the Earth. A solid foundation for the statistical association of large flares and storms was provided by English astronomer, Harold Newton (1893–1985), who, in the late 1920s, surveyed all large flares observed since 1892 and found a significant correlation between those flares and subsequent geomagnetic storms. With the development of radio astronomy in the 1930s, large solar flares came to be associated with more geomagnetic phenomena such as sudden ionospheric disturbances and ground level events.

···

The First Flare

The first step in associating geomagnetic storms with what later became known as solar flares rather than the associated spot regions was the memorable observations on September 1, 1859, by British amateur astronomers Richard Carrington and Richard Hodgson, who independently witnessed a rapid, intense flash of two bright ribbons on the Sun in visible light. This short-lived transient was followed two minutes later by a marked disturbance of the Earth's magnetic field, detected by the geomagnetic instruments at Kew Observatory in London, and then some 17.5 hours later, one of the largest magnetic storms on record occurred. While Carrington was reluctant to suggest a physical connection between the visible event at the Sun and the geomagnetic storm, Balfour Stewart, the Director of Kew Observatory, claimed that they had caught the Sun in the act of producing a terrestrial event. The direct impact on the Earth by an event over 150 million kilometers away had a profound influence on the study of solar-terrestrial relations that reaches to the present day. Over a century and a half after the initial observations, solar and space physicists are revisiting the remarkable event of 1859 in a concerted effort to apply twenty-first-century tools to model its solar and terrestrial effects.

···

The flare observed by Carrington and Hodgson was an example of a relatively rare event, a large white-light flare, in which the optical continuum is enhanced sufficiently over the background solar surface to be visible in contrast. Most flares are not so conspicuous in visible light, reserving their strongest emission for spectral lines such as hydrogen alpha and higher energy EUV and X-ray radiations. Before the advent of space observations, flares were best observed in the hydrogen alpha line and as such were regarded as purely chromospheric phenomena. As observations improved and space astronomy developed, flare phenomena grew to include EUV, X-ray, radio, gamma-ray, and energetic particle emission, all emitted as transient enhancements, either impulsively or gradually, over background levels.

A Hα flare is a short-lived, sudden increase of intensity in the neighborhood of sunspots (see Figure 9.1). Because of the short duration of these events, many early flare observations were hampered by incomplete coverage, making measurements of the time development and brightness determination extremely difficult. The first systematic photometry of solar flares was carried out in the late 1940s by Helen Dodson of the McMath-Hulbert Observatory in Michigan. The flare light curves determined from Hα spectroheliograms demonstrated the complexity of the solar flare phenomena. The most significant feature observed was the immense broadening of the hydrogen emission lines that takes place catastrophically at the time of the flare flash phase. Mervyn Ellison of the Royal Observatory in Edinburgh, Scotland, attributed these great line widths to the presence of an intense beam of energetic electrons accelerated by electric fields in the region of sunspots and pointing to an active role for the magnetic field in the flare process.

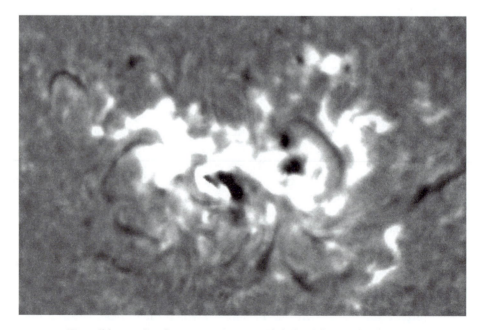

Figure 9.1 Two ribbon solar flares seen in Hα. Global High Resolution Hα Network, operated by the Big Bear Solar Observatory, New Jersey Institute of Technology.

Statistically, the frequency of occurrence of Hα flares was found to vary markedly with the magnetic type of sunspot (see Chapter 5 for a discussion on sunspot classification), increasing as we progress from the simple α spot to the magnetically complex δ spots. Modern theoretical studies invoke the magnetic complexity of the δ configuration to provide the necessary conditions for solar eruptive phenomena. This suggests that an important role is played by the magnetic field in the generation of flare activity (see Chapter 3). Indeed, the first magnetic interpretation for solar flares appeared as early as 1946 when Ronald Giovanelli (1915–1984) of the Commonwealth Solar Observatory in Canberra, Australia, proposed the idea that flares were electrical discharges caused by changing magnetic fields in a high-conductivity medium. The reconfiguration of the magnetic field has been central in understanding flare phenomena ever since, and it is now universally accepted that the whole flare phenomenon is critically governed by the properties of magnetized plasma.

In association with the Hα ribbons it was noted that the expulsion of hydrogen was observed near the peak intensity of the majority of bright flares. These emissions were found to occur in specific directions, usually along nearly vertical trajectories. Whether seen in elevation at the limb or in Doppler shifted Hα against the disk, this expulsion of material exhibited all the characteristics of the well-known eruptive prominences (Figure 9.2). The initial velocity of a mass expulsion was around 500 kms^{-1}, and while

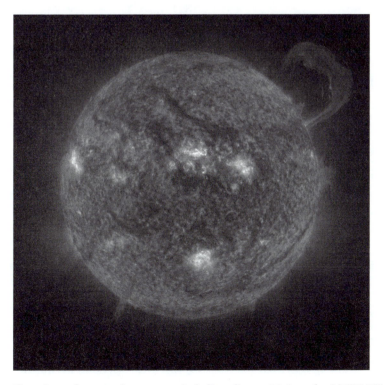

Figure 9.2 Erupting solar prominence seen in helium line at 30.4 nm by SOHO EIT.

its Hα brightness was several times that of normal quiescent prominences, it was still much fainter than the flare emission itself.

Energy Ranges in Flares

Solar flares emit radiation across the whole electromagnetic spectrum, as well as produce fast-moving particles. The physical mechanisms that produce the different wavelengths of the radiation vary significantly, and so the normal subdivisions of the electromagnetic spectrum are subdivided further to provide a more physics-based separation. These are summarized as follows:

Table 9.1. **Electromagnetic radiation from solar flares**

Name	Wavelength Range (nm)	Frequency Range (GHz)	Energy range (eV)	Description
Radio	$>10^5$	$<3 \times 10^3$	$<10^{-2}$	Long-wavelength emission encompassing millimeter wavelengths and above.
Microwave	$10^6 - 10^9$	$3 \times 10^2 - 0.3$	$10^{-3} - 10^{-6}$	Solar flare radio emission signifying the presence of fast-moving electrons traveling along the magnetic field in the location of the flare.
Infrared	$10^3 - 10^5$	–	$1 - 10^{-2}$	Low-energy photons associated with the heating of the solar chromosphere during flares.
White Light	$400 - 700$	–	$3 - 1.8$	This covers the optical emission from flares integrated over all visible wavelengths (different instruments use different wavelength ranges). A flare that is bright in white light is an extremely strong flare.
Hα	656.3 ± 0.05	–	1.9	Red-line emission of historical importance to flare observations. Used to observe chromospheric ribbons of flares, prominence eruptions, and post-flare loop structures.
Ultraviolet (UV)	$50 - 400$	–	$25 - 3$	The UV emission of a flare denotes the upper chromospheric emission at

Table 9.1. (*Continued*)

Name	Wavelength Range (nm)	Frequency Range (GHz)	Energy range (eV)	Description
				temperatures around 10,000 – 50,000 K.
Extreme-ultraviolet (EUV)	6 – 50	–	200 – 25	This wavelength range is sensitive to plasma at temperatures of 1–3 MK and so provides useful pre-flare and post-flare observations. EUV also useful for non-flaring coronal observations.
Soft X-rays (SXR)	0.12 – 6	–	10^4 – 200	Soft X-rays from plasma with temperatures ranging from 2–100 MK provides observations of the hottest temperatures in solar flares as well as coronal observations of the non-flaring solar corona.
Hard X-rays (HXR)	0.006 – 0.12	–	2×10^5 – 10^4	High-energy particles in solar flares interact with the ambient atmosphere to produce radiation at these short wavelengths. Most HXR emission comes from the chromosphere in flares, although more sensitive observations have detected emission from the solar corona in this energy range. HXR emission from flares provides useful diagnostic information on the high energy particles (mostly electrons) accelerated in solar flares.
Gamma Rays	<0.006	–	$>2 \times 10^5$	The highest energy of photons emitted in solar flares signifying the interaction of accelerated protons and ions with the solar atmosphere.

Note: For the larger-wavelength radiation the equivalent frequency is used; as the wavelength decreases, generally energy is used

Table 9.2. Classification of solar flares using Hα emission (Zirin, 1988)

| CORRECTED AREA | | RELATIVE INTENSITY EVALUATION | | |
SQUARE DEGREES	MILLIONTHS OF HEMISPHERE	FAINT (f)	NORMAL (n)	BRILLIANT (b)
<2.06	<100	Sf	Sn	Sb
2.06 – 5.15	100 – 250	1f	1n	1b
5.15 – 12.4	250 – 600	2f	2n	2b
12.4 – 24.7	600 – 1200	3f	3n	3b
>24.7	>1200	4f	4n	4b

Source: Zirin (1988)

The magnitude of solar flares has two historical classifications that are in common usage. One is based on an older optical classification that symbolizes the original chromospheric nature of solar flares. The other utilizes a coronal descriptor based on the X-ray enhancements associated with the heating of the solar atmosphere by flares. The prevalence of the X-ray designation of flares means that the Geostationary Operational Environmental Satellites (GOES) classification is more commonly used in current solar physics.

Optical Classification
Flares are classified on the basis of area at the time of maximum brightness in Hα.

- Importance 0 (Subflare): < = 2.0 hemispheric square degrees
- Importance 1: 2.1 – 5.1 square degrees
- Importance 2: 5.2 – 12.4 square degrees
- Importance 3: 12.5 – 24.7 square degrees
- Importance 4: >24.8 square degrees

(One square degree $= 1.474 \times 10^8$ km$^2 = 48.5$ millionths of the visible solar hemisphere.) A brightness qualifier f, n, or b is generally appended to the importance character to indicate faint, normal, or brilliant (for example, 2b).

X-ray Classification
In this classification, flares are distinguished by the flux of energy (energy per sec. per unit area) at the Earth observed in the 0.1–0.8 nm range by the Geostationary Operational Environmental Satellites, or GOES, operated by the National Oceanic and Atmospheric Administration (NOAA) and better known for their hurricane tracking than their solar monitoring. The system is a decadal system with a letter denomination marking the decade and a number denoting the position within the decade. For instance, a large M-class flare of classification M3.5 has an X-ray flux at the Earth of 3.5 $\times$ 10^{-5} W m^{-2} (see Table 9.3).

Table 9.3. Soft X-ray flare classification based on fluxes observed by GOES satellites

GOES Class	Intensity (Watts m^{-2})
B	10^{-7}
C	10^{-6}
M	10^{-5}
X	10^{-4}

Note: During solar maximum, several large flares occur frequently

Space-based Observations

Our knowledge of the basic physics of solar activity has increased dramatically since we started taking observations from space. Following the success of the rocket program and the flight of the first satellites, space instrumentation developed at an increasing pace. Within the field of solar physics alone, several missions have been flown to study solar activity and, in particular, solar flares. These include NASA missions such as the OSO series of satellites (*Orbiting Solar Observatories*) in the 1960s and early 1970s, *Skylab* in 1973–1974, and the *Solar Maximum Mission* from 1980–1989. More recently, other nations have taken the lead in the study of solar activity (albeit with significant contributions from the United States via NASA). Japanese missions such as *Hinotori* in 1981–1982 and *Yohkoh*, which was launched in 1991–2001, were designed specifically to study solar flares. Predominantly European missions such as *Ulysses* and SOHO, operating currently, targeted the solar wind/interplanetary space and the non-flaring Sun, respectively. The *Ulysses* spacecraft was the first to leave the ecliptic plane since the famous *Voyager* spacecraft. Its orbit takes it over both solar poles to sample the slow and fast solar wind. NASA returned to solar activity studies with the TRACE (*Transition Region and Coronal Explorer*) satellite launched in April 1998 and the RHESSI (*Ramaty High Energy Solar Spectroscopic Imager*) mission launched in February 2002. Both of these missions are still in operation and providing a wealth of data on solar flares. Most recently, the successful deployment of the Japanese-led *Hinode* mission (September 2006) and the NASA STEREO mission (October 2006) have provided unprecedented observations of the response of the solar atmosphere to the energy release associated with solar flares and coronal mass ejections. This fleet of Sun-watching telescopes will be joined in August 2009 by the *Solar Dynamics Observatory*, the first of the NASA *Living with a Star* missions, which will be the first mission to view the entire domain of the Sun where magnetic fields originate and cause solar variability (see Chapter 12).

Developments in ground-based radio observations and the advent of space-based solar physics has completely changed solar flare research to the extent that modern approaches treat flares as primarily a coronal manifestation with enhanced emission of high-temperature radiation, accelerated particle production, mass ejection, and rapid morphological changes.

Most flares exhibit significant brightness enhancements in the EUV, soft X-ray (SXR) and hard X-ray (HXR) energy ranges, with the EUV and SXR signifying the substantial heating of the solar atmosphere to temperatures as high as 40 million Kelvin, and the HXR emission indicating the presence of particles with energies orders of magnitude higher than the thermal

Figure 9.3 Postflare loop structures from TRACE EUV observations. Courtesy of NASA.

energy of the background plasma. Figure 9.3 shows the post-flare enhancement in the EUV. The post-flare magnetic structure (an arcade of magnetic loops) is clearly evident. While the energy content of the flare is fairly evenly distributed between the accelerated particles and the heated plasma, the hard X-ray emission provides a direct signature of the energy release process associated with solar flares, indicating the interaction of flare-accelerated particles with the ambient solar atmosphere—most significantly, the chromosphere that is dense enough to "stop" a high energy particle (see sidebar).

Hard X-ray radiation (photon energy >10 keV) from solar flares was first observed as far back as 1959 by Lawrence E. Peterson and John Randolph Winckler using a balloon-borne X-ray detector. Subsequently, rocket, balloon, and satellite observations improved our knowledge of the hard X-ray emission that was thought to be *bremsstrahlung radiation* from energetic electrons interacting with the ambient solar atmosphere (see sidebar).

The most direct observational evidence for the acceleration of particles at the Sun came with the first signs of radio emission produced by fast electrons interacting with solar plasma and magnetic fields. This was subsequently followed by the discovery of radiations at higher energies (e.g., hard X-rays and gamma-rays), implying particles with energies of several tens of MeV, and direct measurement of energetic electrons and ions in interplanetary space. These energetic radiations are all presumed to be generated by

Bremsstrahlung Radiation

Particles, accelerated in the solar corona as a result of the initiation of a flare, travel at high velocity through the solar atmosphere. These particles, typically electrons or protons, have an electric charge and will feel the effects of the charged particles that make up the atmosphere. Typically, the accelerated particle is an electron with a negative charge of magnitude $q = 1.6 \times 10^{-19}$ Coulombs, mass $m_e = 9.8 \times 10^{-31}$ kg, and velocity v. A typical velocity for a hard X-ray producing electron is about 60,000 km/s (or 20% of the speed of light). This accelerated electron interacts with a proton in the solar atmosphere. This target proton has a mass $m_p = 1.67 \times 10^{-27}$ kg, a positive charge of magnitude $q = 1.6 \times 10^{-19}$ Coulombs and is assumed to have zero velocity (a proton in the solar atmosphere will have a velocity associated with the temperature of the plasma, which is typically 30 km/s, and so is effectively stationary compared to the accelerated electron). As the electron gets closer to the proton, it starts to feel the electrostatic attraction and is pulled towards the proton, away from its initial trajectory. The path of the electron bends around the proton, giving off a photon in the process and resulting in the electron traveling in a different direction from when it started and with slightly less energy. Because the electron slows down as a result of this interaction, it is given the German name *bremsstrahlung* or "braking radiation." These interactions are repeated many times for every accelerated electron (typically $10^{34\text{-}36}$ electrons are accelerated every second in a flare) and with many of the ambient target particles (in the chromosphere there are about 10^{19} protons per cubic meter). The sum total of all of these particle interactions is to produce the hard X-ray signatures associated with solar flares (Figure 9.4).

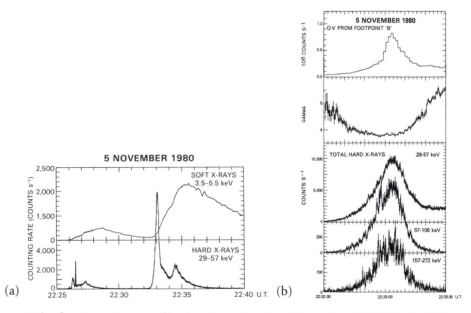

Figure 9.4 Solar flare X-ray time profiles for a large flare from November 5, 1980. (a) Soft X-rays are smoothly varying, while hard X-rays show impulsive bursty structure. (b) Hard X-rays show different bursty structure at different energies. With kind permission from Springer Science + Business Media: Solar Physics, "Solar Hard X-Ray Bursts" (ISSN 0038-0938), vol. 100, October 1985, p. 473–474, Brian Dennis, Figures 5 and 6.

the interaction of energetic particles accelerated in the solar flare that subsequently interact with the ambient solar medium. Many populations of energetic (i.e., nonthermal) particles created by solar flares have been identified covering several decades in energy. These populations include deka-keV electrons (10 –100 keV) responsible for the bulk of the flare hard X-ray emission, relativistic electrons (>10 MeV) contributing to the gamma-ray continuum, accelerated ions (10–100 MeV) that generate the rich nuclear de-excitation gamma-ray line spectrum, >100 MeV ions that result in production of energetic pions in the solar chromosphere, and high-energy neutrons that result from the interaction of energetic protons and alpha particles in the flaring solar atmosphere. To varying extents, particle acceleration appears to occur in all flares, including relatively minor ones. Knowledge of the particle numbers, energy distributions, and relative abundances provide the key to unlocking the secrets of the flare energy release (see sidebar).

The direct relationship between the observed amount of hard X-ray emission of photons with energy above 20 keV and the energetic particles that produced them place significant constraints on the processes driving solar flares. In particular, the comparison of the total numbers of particles and energy required to explain the observed emission with the energy available from the magnetic field puts the difficulty of understanding the flare process into sharp relief.

Total energy in particles over a 100s flare:	$\sim 10^{25}$ Joules
Available magnetic energy:	$\sim 10^{25}$ Joules

(proportional to square of the magnetic field times the volume)

This implies that a large fraction of the available magnetic energy is converted into accelerated particles.

Total number of electrons required to produce observed hard X-rays:	$10^{36\text{-}37}$ s^{-1}
Total number of particles available in flare volume:	10^{37}

(proportional to density times volume)

This implies that the whole coronal structure taking part in the flare has to be emptied (and refilled) every second for the duration of the flare.

These constraints are quite extreme given that the conversion of magnetic energy into particle acceleration and heating is best achieved in small low-density volumes and that the acceleration efficiency is, at best, only 1%–10%. In other words, the fact that the amount of energy released and the number of particles involved require the whole volume and effectively 100% efficient acceleration is a major issue in understanding the energy conversion process that drives solar flares. Solar flare scientists are exploring all aspects of this problem to determine whether there is more free energy available in the magnetic field than previously considered, whether larger volumes are involved, or whether the inference of the particle numbers and energies from the observed hard X-ray photon radiation is correct.

Solar flares are among the most energetic phenomena in the solar system and arguably in the universe—a solar flare has a luminosity to mass ratio, L/M, of magnitude 10^{38}, significantly higher than cosmic explosions such as blazars (related to supermassive black holes at the centers of galaxies) with L/M $\sim 10^{33}$, and surprisingly close, given their size, to gamma-ray bursts, which have L/M $\sim 10^{45}$. Coupled to the fact that flares are responsible for the production of particles with energies as high as 0.1 to 1 GeV makes solar flares important for understanding energetic processes in astrophysics. Closer to home, the impact of solar flare phenomena, from increasing the radiation content of interplanetary space to energizing the Earth's atmosphere, makes them an extremely important component of space weather in conjunction with their frequently associated counterparts, coronal mass ejections.

CORONAL MASS EJECTIONS

It had long been known that clouds of particles were expulsed from the Sun at the time of large solar flares and that these particles resulted in the dramatic response of the Earth's magnetic environment in the form of geomagnetic storms. The problem was that this plasma cloud could not be detected directly on its journey from the Sun to the Earth, despite some ingenious ideas and observations. Originally it was thought that the source of these particles was assumed to be a blast wave, or shock wave, generated by the flare, which would reach the Earth in two to three days. Detailed observations subsequently suggested that the shocks could not be generated by the waves themselves, but no other explanation was put forward. Then, in 1973, the *Skylab* space station was launched, and the source of the mysterious shocks was discovered.

Among the instruments flown on board *Skylab* was the most sensitive coronagraph built up to that time. Indications of large, transient disturbances traveling through the Sun's outer corona had been noted in solar radio records and found in coronagraph observations from earlier unmanned spacecraft (most notably OSO-7). However, the *Skylab* coronagraph observed in such detail that a new solar phenomenon was discovered—the coronal mass ejection. The astronauts on the space station witnessed gargantuan loops rushing outwards from the Sun at fantastic speeds, and astronomers on the ground were elated by the first detailed pictures of the expulsion of erupting material that ultimately grew to be bigger than the disk of the Sun.

Figure 9.5 CME observed by SOHO LASCO telescopes showing three-part structure (bottom left). Courtesy of NASA.

Coronal mass ejections have been a topic of extensive study since their discovery and have proved to be scientifically interesting in many different ways. When viewed, the edge on the coronal mass ejection has a distinctive three-part structure: a bright core that is the remnant of an eruptive prominence; a large, dark, lower-density surrounding cavity; and an outer, diffuse leading edge having the projected shape of a closed loop with its legs fixed on the Sun (Figure 9.5).

A key feature of the CMEs is their transport through interplanetary space and subsequent interaction with the Earth. A typical mass ejection contains approximately 7×10^{11} kg of material, and the outward flow of this material drives pressure waves into the surrounding corona and solar wind. When the outward speed is sufficiently high, the ejections produce shock waves in the solar wind far from the Sun (see sidebars). The advent of CMEs as the generators of the interplanetary shocks supplanted solar flares as the primary source of geo-effective energetic particles. Flares are not required to produce a CME and are arguably only secondary phenomena when they do occur together. Solar flares do, however, have terrestrial consequences, as evidenced by the detection of the sudden ionospheric disturbances and ground-level enhancements discussed earlier; such terrestrial effects were major disruptors of radio communications prior to the satellite era. Recent observations from the coronagraphs on board the *SOHO* spacecraft (launched in 1995 and still in operation today) have introduced several new aspects that have added significantly to our knowledge of CMEs:

1. Many are accompanied by a global response of the solar corona: disruptions circling the entire Sun at a height of 1–2 solar radii have been observed to accompany large CMEs;

2. Many show continuous acceleration all the way out to the edge of the instrument field of view—whatever is driving these disturbances acts on the CME mass over a distance of millions of kilometers;

3. Disconnection is a frequent occurrence—frequently the mass associated with CMEs is seen to be completed devoid of any magnetic connection back to the Sun, a key feature in understanding how CMEs get started;

4. CMEs are occurring more frequently than had been expected at solar minimum—this provides important clues to the inner workings of solar variability; and

5. CMEs undergo extensive internal evolution as they move outward.

Coronagraphs

The discovery of CMEs relied on a unique instrument called a *coronagraph* developed by French astronomer Bernard Lyot (1897–1952) in 1930. This instrument effectively creates a fake eclipse by blocking off the solar disk and allowing the much fainter corona to come into sharp relief. The Sun is blocked using a thin metallic disk known as an *occulting disk*. CMEs are detected by Thompson scattered light (see Chapter 7) where light from the disk of the Sun is scattered by electrons in the corona. The passage of the CME through the corona locally enhances the density of electrons, thereby increasing the scattering and making it stand out against the background corona. Coronagraphs have improved substantially over the intervening eighty years with increasing sensitivity, so now some space-based coronagraphs on SOHO and STEREO are sensitive to small changes in the electron density out to distances greater than 30 solar radii (20 million km). More important, is the height above the solar surface of the inner edge of the coronagraph. The brightness of the Sun makes it difficult to observe low in the atmosphere since any slight jiggle of the spacecraft or error in the alignment of the mirror and the occulting disk can let light from the Sun into the telescope. This can wash out the signal from the faint corona and in some cases even damage the telescope. Modern coronagraphs can get as close as 0.14 of a solar radius (100,000 km).

Fast CMEs are a major source of space weather as a result of an interesting interaction with the solar wind. As pointed out in Chapter 8, the slow solar wind, relevant to the ecliptic plane, propagates from the Sun at a speed of approximately 400 km/s. Fast CMEs typically travel in excess of 500 km/s and even as high as 2500 km/s. This means that when a fast CME takes off, it is traveling faster than the solar wind through which it is passing. As a result, the CME effectively plows into the solar wind plasma to create shock waves that travel ahead of the propagating CME. The faster the CME, the stronger the shock waves it produces. The shock/CME system then carries the disturbance outwards into interplanetary space—ahead of the shock the solar wind is unperturbed, traveling at its normal undisturbed velocity. The transition from shocked plasma to solar wind plasma is quite dramatic, with the speed and density dropping precipitously between the two. As the CME expands outwards, it produces a broad region of strong magnetic field between the CME and the shock. The combination of strong magnetic field and high flow speed provide ideal conditions for accelerating particles. The electrons and protons in the plasma surrounding the shock can gain energy from the magnetic fields and be accelerated to very high energies. These very energetic particles can then escape ahead of the shock and propagate along the solar wind magnetic field lines (the interplanetary magnetic field discussed in Chapter 8) and interact with the Earth's magnetosphere. The sheer extent of the shock ahead of the CME means that, unless the CME occurs far around on the side of the Sun away from the Earth, the Earth will be magnetically connected to some part of the shock and therefore be affected by the accelerated ions and electrons. This is very different from the situation with the flare-accelerated particles that can only be connected to the Earth along the Parker spiral. Because of this, CMEs are regarded as a far more important source of geoeffective activity. CMEs that have speeds slower than the ambient solar wind speed do not produce shocks.

THE SOURCE OF ERUPTIVE PHENOMENA

One of the outstanding questions in understanding flares, CMEs, and their relationship is how they are initiated. While they may occur separately, there is increasing evidence that a significant energy release is often accompanied by an eruption of material and a localized heating. A commonly observed phenomenon that seems to be intricately linked with this energy release is the solar prominence (or solar filament)—prominences and filaments are the same thing viewed differently (see Chapter 7). The first observations of prominences on the limb, outside of a solar eclipse, were those of British astronomer Joseph Lockyer and French astronomer Jules Janssen, in 1868, using spectroscopes to measure prominence emission lines in full daylight. It was from observations of prominences that the element helium was discovered on the Sun. It wasn't until 1903, with the aid of photographic techniques that American scientists George Ellery Hale and Ferdinand Ellerman discovered that the dark filaments observed in spectroheliograms were merely the projection of prominences on the disk (see sidebar).

The relationship between solar flares and prominences goes back several decades to the *disparition brusques* phenomena catalogued in the late 1940s by researchers at Meudon Observatory in France. Nearly all low-latitude filaments were found to disappear temporarily at least once during their lifetime, and this disappearance was frequently related to the occurrence of a solar flare. However, despite the fact that solar prominences had been known for several hundred years, and were easily observed, they were not thought to play a role in geomagnetic storms. A relationship was suggested

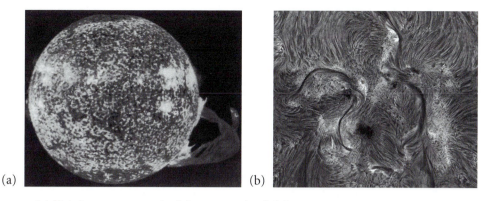

Figure 9.6 (a) This image was acquired from NASA's *Skylab* space station on December 19, 1973. It shows one of the most spectacular solar flares ever recorded, propelled by magnetic forces, lifting off from the Sun. It spans more than 588,000 km (365,000 miles) of the solar surface. In this photograph, the solar poles are distinguished by a relative absence of supergranulation network, and a much darker tone than the central portions of the disk. Courtesy of NASA. (b) Large solar filament seen in H-alpha by the Dutch Open Telescope.

by Harold Newton and William Greaves in 1928, but Hale disagreed, pointing out three years later that erupting prominences generally fall back to the Sun. The connection between prominence eruptions and geomagnetic storms was not fully appreciated until the work of JoAnn Joselyn and Patrick McIntosh in 1981. The flare was regarded as the primary agent for geoeffectiveness, although it was clear that there was a strong correspondence between filament activity and flaring in solar active regions. German solar physicist Karl Otto Kiepenheuer (1910–1975) demonstrated that prominences were observed to rise into the corona with increasing velocity that may eventually exceed the velocity of escape, thereby explaining the sudden disappearances. This process was studied in detail, with the conclusion that the ejected plasma is accelerated as it rises. Such studies were the precursors to present-day investigations into the relationship between filament eruptions and flares, and preceded by as much as three decades the discovery of coronal mass ejections.

Today it is generally recognized that among all manifestations of coronal activity, statistically the greatest correlation is between CMEs and eruptive prominences. For the larger, and many of the smaller, CMEs, this also implies a relationship between eruptive prominences and flares. An important factor is the height that prominences reach—the greater the height, the more probable a CME will result. For example, many studies show that prominences that attain a height greater than 70,000 km have a greater probability of producing a CME. However, it should be borne in mind that the association of erupting prominences and the flare/CME may not imply causation, but all three phenomena may be different observational manifestations of the same coronal disruption.

Discovery of Chromosphere

The solar chromosphere is too faint to be seen against the bright glare of the solar surface and was only visible, historically, during the short periods near the beginning and end of a total eclipse. The earliest recordings of the chromosphere can be traced back to the eclipse observations of 1706, when English sea captain Henry Stannyan reported seeing a blood-red streak that lasted six or seven seconds at the western limb of the Sun. Similar observations were reported from the eclipse of 1715 by Edmund Halley and Jacques Eugene d'Allomville, Chevalier de Louville. This vivid and readily observable layer was dubbed the *sierra* by British astronomer Sir George Airy (1801–1892) after his observations of the great European eclipse of 1842, but the name *chromosphere*, or *color-sphere*, was adopted by astronomers after being introduced by British scientists Sir Edward Frankland (1825–1899) and Sir Norman Lockyer after the eclipse 1869 as a phrase representing the redness of this predominantly hydrogen layer.

Transition Region

The solar transition region marks the sharp transition from chromospheric temperatures of around 12,000 K to coronal temperatures in excess of a few hundred thousand degrees. The need for such a transition was evident from the spectroscopic temperature measurements of the chromosphere and corona in the 1940s and 1950s, although observations could not measure the intermediate temperatures at that time. The concept of a distinct layer, or transition region, came about from a number of theoretical calculations that were developed to examine how heat was transferred through and between the different layers of the solar atmosphere. Several models developed between 1946 and 1956 encapsulated a sharp transition between the chromosphere and corona, but it was a theoretical argument given in 1948 by Australian scientist Ronald Giovanelli (1915–1984) that suggested the existence of a transition region, a region of high temperature gradient, where the main agent of heat transfer changes from convection in the chromosphere to thermal conduction in the corona. Giovanelli further suggested that this region be identified as the boundary between the chromosphere and corona.

PROMINENCES AND FILAMENTS

One of the most prominent features observed at chromospheric temperatures is the solar prominence. A prominence is a large-scale, sometimes several hundred thousands kilometers long and tens of thousands of kilometers high, chunk of the chromosphere embedded in the solar corona. Pictures of the Sun in the intense hydrogen line known as Hα often show these dramatic structures seemingly floating in the solar atmosphere but tethered to the surface, like a hot air balloon ready for takeoff. While the surrounding corona is around a million degrees and has a density around 100 trillion (10^{14}) particles per cubic meter, the material in the prominence has a temperature around 10,000 degrees and a density as high as 10^{18} particles per cubic meter. The density and temperature of the prominence

THE SOLAR STORM

While living near a star has many advantages, the Sun is not a quiet neighbor, and Earth's location puts it in the path of explosive solar storms. The plasma cloud ejected during a solar storm can travel the vast distance from the Sun and reach Earth in as little as one to four days. At one time, the only evidence of the arrival of this cloud was the spectacular display of color it would produce in the night sky—the northern or southern lights. However, with our increased reliance on technology, solar storms now have a much more disruptive effect on everyday lives.

Space weather disturbances can disrupt satellite operations, and interfere with electronic and radio communications, impacting our lives and inconveniencing us for prolonged periods of time. One of the more destructive effects of solar storms is their ability to short-circuit power grids, leaving thousands of people without electricity. Increased radiation from solar storms can also be harmful to astronauts in orbit. To improve predictions of what is called space weather, scientists work to better understand the causes of solar storms through careful observations of the Sun leading up to a storm, and of the storm itself. The life of a solar storm is interesting and exciting and can be told through a careful examination of such observations.

There is still much we do not know about solar eruptive events, and it is clear that flares and CMEs will continue to inspire solar research for a long time to come. Of particular interest is the ability to predict these events and their potential terrestrial impact. Predicting the behavior of something as physically complex as the Sun is, like terrestrial weather, an extremely difficult endeavor. However, the array of ground- and space-based observatories coupled to improvements in instrumentation and physical models of the Sun-Earth system will significantly improve our chances of attaining this goal.

....

Story of a Solar Storm

In spring 2001, an extremely large sunspot group surfaced on the Sun. As the colossal dark region crossed the solar disk, it grew to more than thirteen times the size of Earth. Known as active region number 9393 (the 9393rd region to be observed since counting officially began in 1973), this giant was the largest sunspot group to appear in ten years! Active region 9393 proved to be not only large in size, but persistent in nature. The region remained strong for three complete solar rotations, each lasting twenty-seven days, revealing itself in March, April, and May 2001.

On April 2, 2001, at 5:51 PM EDT, active region 9393 released a major solar flare. The flare occurred off the northwest limb of the Sun and was quickly determined to be one of the largest flares on record. Fortunately, the accompanying eruption, a coronal mass ejection, was not pointed directly at Earth, helping to limit the damaging impact of the storm.

Several days later, a large influx of energetic particles generated by the April 2 solar storm arrived at Earth, producing beautiful auroral displays that were seen as far south as Mexico. This particular storm did little damage to the Earth's electrical systems, with no major satellite disruptions or power-grid failures reported.

....

RECOMMENDED READING

Carlowicz, Michael, and Ramon Lopez. *Storms from the Sun: The Emerging Science of Space Weather*. Washington, DC: National Academies Press, 2002.

Zirin, Harold. *Astrophysics of the Sun*. Cambridge: Cambridge University Press, 1988.

WEB SITES

Big Bear Solar Observatory: http://www.bbso.njit.edu.

The First Flare: http://science.nasa.gov/headlines/y2008/06may_carringtonflare.htm.

GOES X-ray Imager: http://sxi.ngdc.noaa.gov.

RHESSI Flares: http://hesperia.gsfc.nasa.gov/hessi/flares.htm.

SHINE: www.shinecon.org.

Skylab: http://www-pao.ksc.nasa.gov/history/skylab/skylab.htm.

10

Space Weather

The previous chapter focused on the most energetic of solar phenomena, namely the various components that make up a solar storm: solar flares and coronal mass ejections (CMEs). Solar flares emit electromagnetic radiation over an extensive range of wavelengths, but notably in the more dangerous high-energy ranges from extreme ultraviolet, to X-ray and gamma-ray. In addition, they emit highly energetic particles, mostly protons and electrons. Fast CMEs plow into the slower-moving solar wind to create shocks that extend over large distances and propagate away from the Sun. These shocks are regions where electrons, protons, and heavier ions can be accelerated, sometimes up to speeds close to the speed of light. The enhanced electromagnetic emission and accelerated particles from flares and CMEs are all superimposed on a background solar wind comprised of ions and electrons traveling at 400 km/s to 800 km/s (Chapter 8). On Earth, we are protected from much of this harmful radiation by the presence of both an atmosphere and a strong magnetic field. However, as we begin to explore beyond the Earth's magnetosphere, placing resources, hardware, and people in space, we must confront the potential hazardous consequences of the Sun's electromagnetic and particle radiation, in particular, the large increases in this radiation produced by solar storms. The challenge of radiation safety for crew and equipment is crucial to NASA's new visions of long-duration manned presence across the solar system. Long-duration sojourns in space by astronauts, as would be required for the proposed Lunar-Base and Mars Exploration Missions, will be at the mercy of solar radiation for the duration of their mission and will no longer be able to rely on the low exposures of the earlier week-long Apollo lunar excursions.

Closer to home, solar radiation can influence the performance and reliability of spaceborne electronics and hardware and adversely affect the safety of humans in space. For example, solar radio bursts can produce excess noise in wireless communications, while solar particle radiation can cause a range of effects, including increased degradation of solar panels, single event upset damage to sensitive electronics, anomalous electrical charging of spacecraft, and the increased exposure of astronauts to harmful radiation. On the Earth, we are protected from much of this harmful radiation by the presence of both an atmosphere and a strong magnetic field. However, as we have continued to populate the space around the Earth with satellites and space stations, establishing both hardware and crews in space, we are increasingly forced to confront the potential hazardous consequences of the Sun's output.

Since 1995, the National Space Weather Program, an initiative spanning several government agencies including NASA, the National Science Foundation, and the Departments of Commerce, Defense, Energy, and the Interior, has sought to address the impact of the Sun on space (and terrestrial) resources via an integrated and targeted approach to solar observation, modeling, event prediction, and forecasting. These tools employ a wealth of solar monitoring instrumentation, computer modeling, and human experience to safeguard the various resources that we have placed within the Earth's space environment. Current space weather efforts concentrate on the potential for solar storms to be geo-effective, namely how likely an event on the Sun is to significantly affect the Earth's magnetic environment. Future proposed space endeavors include the establishment of a permanent base on the Moon and the human exploration of Mars, and the tools developed for the near-Earth space environment need to be modified for these new regimes—long stays on the Moon's surface with little or no natural protection from solar radiation, and extended voyages to Mars, lasting as long as twenty-six months, with full exposure to the Sun the whole way.

The interaction of the solar wind and solar disturbances with the Earth's magnetosphere presents its own set of unique problems and challenges for our understanding of space weather and, in particular, our ability to forecast impending geomagnetic storms. For instance, the orientation of the magnetic field embedded in the incoming plasma is critical to how geo-effective the event is. For missions to the Moon, Mars, or during long-duration transit, the relative luxury of protection by the Earth's magnetospheric shield is not available. Continued exposure to potentially harmful solar radiation is a key factor in the planning and preparation for missions to the Moon and Mars involving manned spacecraft.

Solar energetic particles (SEP) with energies as high as 1 GeV per nucleon are produced by the Sun in association with large flares and CMEs. Very large particle events with large fluxes of high-energy particles are called solar proton events (SPE). An SPE has a precise definition such that the flux of protons with energy above 10 MeV must be greater than 10 particles

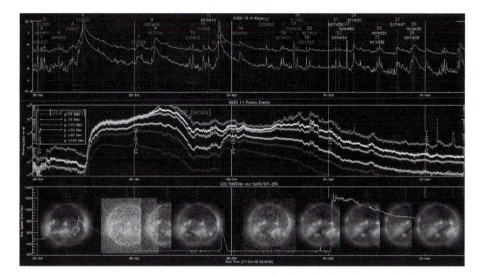

Figure 10.1 Summary figure from the large Halloween storms of 2003. Courtesy of www.lmsal.com. Top panel shows the GOES X-ray flux over the period from October 28, 2003, to November 1, 2003. Middle panel shows the proton fluxes observed at GOES over the same period. Bottom panel shows SOHO/EIT 284Å images with the ACE/SWEPAM measurements of the bulk solar wind speed superimposed. Source: Solar X-ray images are from the Yohkoh mission of ISAS, Japan. The X-ray telescope was prepared by the Lockheed-Martin Solar and Astrophysics Laboratory, the National Astronomical Observatory of Japan, and the University of Tokyo, with the support of NASA and ISAS.

$cm^{-2} s^{-1} sr^{-1}$ at the Earth. Even though SPEs do not occur very frequently, the enormous energy of the particles can result in some of the most hazardous impacts on satellites and spacecraft in low-Earth orbit, as well as significantly affect the Earth's atmosphere. On the even rarer occasions when multiple SPEs occur in a relatively short time, the radiation dose can increase substantially, increasing the danger to astronauts and hardware. Figure 10.1 shows a recent period of multiple solar flare activity resulting in SPE level particle fluxes lasting several days. Analyses of large events have clearly demonstrated that SPEs have the potential to be hazardous to astronauts over a range of timescales, increasing both total accumulated radiation dose and enhancing dose rates. Since radiation exposure has permanent effects, determination of potential long-term exposure can allow for precise mission planning to minimize unnecessary exposure.

The challenge of radiation safety for crew and equipment is crucial to NASA's new vision of long duration and, ultimately, continuous manned presence in the solar system. Long-duration missions outside of the Earth's magnetosphere, such as the proposed lunar-base and Mars exploration missions, will no longer be able to rely on the low integrated exposures of earlier week-long lunar excursions. Meeting this challenge will require the development of effective solar flare and CME prediction systems as a critical component in mitigating the overall radiation exposure. Prediction of solar flares currently relies upon the identification of local pre-flare signatures within solar active regions, in particular the strength, complexity, and dynamics of the magnetic field. CMEs, in contrast, span significantly larger spatial scales than flares and are often not causally associated with flares. While active regions display several signatures associated with CME production, none of these signatures have been shown to provide an accurate prediction of when a CME will occur, nor what its properties will be.

It is important to point out that while CME and flare prediction can be interrelated, for most large events they differ substantially in many aspects. First, flares and CMEs can, and do, occur in isolation. Second, it is not clear that when they do occur together that there is a causal connection. Third, the particle production processes involved may be quite different. Fourth, flare and CME-induced particles require access to open magnetic field along which to propagate out into the heliosphere. Flares tend to be localized, while CME-generated particles have access to open magnetic field across a wide range of solar longitudes.

Historically, automated prediction systems have been limited to the use of solar photospheric observations, such as the McIntosh spot classification system (see Chapter 5), with only 10%–20% reliability in the case of flare prediction. Currently there are no operational automated CME prediction systems. However, the recent revolution in solar physics observations, with routine high cadence monitoring of all layers of the solar atmosphere provides a wealth of long-term archival data ripe for statistical study. Despite these advances, there has been no systematic attempt to incorporate all available observational datasets and computational modeling results into a single forecasting framework.

The particle and electromagnetic (photon) radiation, produced by solar flares and CMEs, result in an array of consequences at Earth. The high-energy particles, in particular, can constitute a radiation hazard for spacecraft, astronauts, and the crews and passengers of high-altitude polar air flights. The associated enhancements in the high-energy particle content of the upper atmosphere of the Earth at high latitudes can lead to spectacular auroral displays, with the largest storms generating aurora as far south as Houston, Texas, and southern California. The high-energy photons, such as those at EUV and soft X-ray energies, pass unimpeded through the Earth's magnetic field and are absorbed by the upper atmosphere. As a result, the atmosphere becomes more ionized, and the high-energy photons effectively "kick" electrons off the atmospheric atoms. This, in turn, changes the properties of the upper atmosphere sufficiently to cause interference of short-wave radio communications that rely on reflection off the ionosphere, as well as a general expansion of the upper atmosphere leading to an increased drag on low-Earth orbiting satellites and a shortening of the orbital lifetime of these satellites.

One of the major space weather impacts associated with solar flares and coronal mass ejections is the proton storm comprising a large increase in the amount of high-energy ions, mostly protons, in Earth's near-space environment. Fast-moving protons from the Sun are constantly bombarding the Earth in the form of the solar wind (Chapter 8), but typically the energy of these particles is low enough that they do not penetrate far into the Earth's magnetosphere. The fast proton population is significantly enhanced when a solar flare occurs. The energy of the flare-produced protons can exceed hundreds of millions of electron-Volts (usually written as 100 MeV), and as a result they can force their way deep into the Earth's magnetosphere, reaching the ionosphere and increasing the ionization of atmospheric particles. Only flares that are magnetically connected to the Earth can lead to solar proton storms (see Parker Spiral discussion in Chapter 8). Typically, the protons enter the atmosphere at high latitudes, above the poles, guided there by the Earth's magnetic field. Particles in a typical proton storm can reach the Earth within a few hours to days of the detection of the flare in solar imagers (the photons reach the Earth in 8 minutes 19 seconds, the light travel time). However, a large flare that occurred on January 20, 2005, produced the largest proton storm on record, with the remarkable result that the protons reached the Earth around fifteen minutes after the detection of the photons, suggesting that these particles were traveling at around one-third the speed of light. This raises major concerns for human space activities, particularly the proposed exploration of the Moon and Mars (see below), since warnings of the arrival of the largest events may be tens of minutes, too short for an astronaut on a spacewalk to reach the safety of the spacecraft or shelter. The high energy of the individual particles in a solar proton storm (upwards of 100 MeV), means each particle is moving fast enough to penetrate up to 11 cms of water! A spacesuit would afford little protection against such an onslaught. This chapter focuses on space weather and the effects that solar activity has on the Earth.

GEOMAGNETIC STORMS

A geomagnetic storm is a short-lived disturbance in the Earth's magnetosphere caused by the direct interaction of energetic particles from the Sun and the magnetic environment of the Earth. There are generally two types of geomagnetic storm: those associated with the solar wind (Chapter 8), and those associated with transient solar phenomena such as solar flares and coronal mass ejections (Chapter 9). Geomagnetic storms show up in a variety of ways. The solar wind not only carries energetic particles from the Sun, but also magnetic field, and the interaction of the solar magnetic field with that of the Earth generates an array of electromagnetic phenomena at the Earth. Since the orientation of the Earth's magnetic field is northward, that is, the field points outwards from the south pole and inwards at the north pole (see Chapter 3), solar magnetic field of the opposite polarity, that is, southward, is most effective at interacting with the magnetosphere. When the right conditions are met, magnetic and particle energy from the solar wind is rapidly injected into the Earth's magnetosphere. This can increase the intensity and extent of the aurora at both poles, significantly enhance the electrical currents in the ionosphere that can generate problems on the ground (see next section), temporarily enhance the strength of the magnetic field at high latitudes, and provide a radiation hazard to high-altitude polar flights, Earth-orbiting spacecraft, and astronauts in the International Space Station. During large solar flare or CME-induced geomagnetic storms, all of these effects are magnified.

Geomagnetic Indices

The interaction of the solar wind and solar storm plasma generates changes in the Earth's electrical current systems, which, in turn, result in variations in the magnetic field. Fluctuations in the solar wind are therefore reflected in the electrical and magnetic environment of the Earth, and it is these fluctuations that are generally characterized by a commonly used geomagnetic index, denoted the *Ap index*. Interactions within the magnetosphere, and between the magnetosphere and the ionosphere, also contribute to variation in the geomagnetic indices.

There are several geomagnetic indices in current use, but geomagnetic storms are generally defined in terms of their Ap value. The Ap index is defined in terms of another index known as the *Kp index*. The Kp index is a planetary-wide measurement of variations in the magnetic field relative to a "standard" quiet day value, averaged over every three-hour period during the day, and over thirteen geomagnetic observatories lying between 44 and 60 degrees latitude north or south.

The Ap index is the planetary average value of the *daily* averages for each observatory measuring the geomagnetic activity. In other words, for each observatory, a daily average of the eight three-hour magnetic measurements is collated (this is known as the *A index*), and then all of the averages from the thirteen observatories are averaged to get the planet-wide Ap index. The level of geomagnetic activity is broken into four main types: Quiet (Ap < 8), Unsettled ($8 \leq Ap \leq 15$), Active ($15 < Ap < 30$), and Storm ($Ap \geq 30$). The storm category is further divided into Minor Storm ($30 \leq Ap < 50$), Major Storm ($50 \leq Ap < 100$), and Severe Storm ($Ap \geq 100$).

Geomagnetic storms provided one of the earliest signs that there was a strong connection between events on the Sun and phenomena at the Earth, above and beyond the usual heat and light connection. The first solar flare detected by Richard Carrington and Richard Hodgson in 1859 had a clear response at the Earth, but because of the unique nature of that event, very little was made about possible connections; Carrington, himself, was wary of reading too much into the discussion of solar events influencing the Earth, making his famous statement that "One swallow does not make a summer" (see Chapter 9). Although there was much discussion on the topic in the preceding decades, the story only really began to take shape some sixty-eight years later, in 1927, when British astronomers Charles Chree and James Stagg detected a marked twenty-seven-day recurrence in geomagnetic activity. This twenty-seven-day periodicity, matching solar rotation, was convincing evidence for a repeatable source of magnetic storms on the Sun. The problem was that no one at that time could identify the actual location on the Sun from which these storm particles originated. Statistical studies showed that they were not associated with sunspots or other regular features that could be observed. As a consequence, the mysterious source regions were named M-regions by German scientist Julius Bartels. Ultimately, the M-regions were identified to be coronal holes, which were only discovered in the 1970s when X-ray observations from space became available (see Chapters 3, 7, and 8). The source of the recurring geomagnetic activity was the fast solar wind streams associated with coronal holes that extend down from the Sun's polar regions.

Colors of the Aurora

If you have ever seen the auroras (the northern or southern lights), you will generally have seen them as predominantly green in color. However, on a particularly clear night when the aurora is active, you might also see a wide array of colors ranging from red and pink to blue and violet. The source of these varied colors is the interaction of the charged particles from the Sun with the different elements and compounds, mostly oxygen and nitrogen, that make up the Earth's atmosphere. When the charged particles from the Sun collide with the atoms and molecules of the atmosphere,

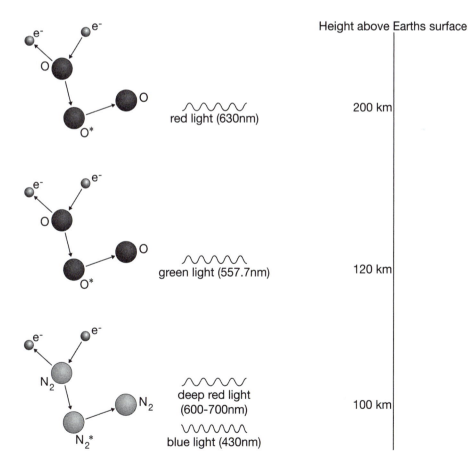

Figure 10.2 Schematic showing the production of the auroral colors. Illustration by Jeff Dixon.

they impart energy to the atmospheric atoms, which then release the energy in the form of light. Different excited atoms give off different colors (see the discussion on emission lines in Chapter 2).

The strong, green light comes from oxygen atoms at heights of 120 to 180 km. The rarer red northern lights also come from oxygen atoms, but from much higher in the atmosphere ($\sim$200 km). Blue and violet light occur as a result of molecular nitrogen mostly below 120 km. When the Sun is particularly active, red aurora can occur at altitudes of around 100 km. The color of the aurora depends on the combination of the particular atmospheric gas, its electrical state, and the energy of the incoming particle.

THE GREAT STORM OF 1989

In early March 1989, a large sunspot, known as region 5395, rotated around onto the visible disk of the Sun. In the days preceding the appearance of this sunspot, observers had noted intense solar activity coming from behind the eastern edge of the Sun—large solar flare enhancements in X-rays and a number of coronal mass ejections. Thus, sunspot 5395 was anticipated and expected to be a fairly large and complex sunspot, but nothing had prepared the solar physicists for what they actually saw. On March 6, the portion of the solar surface containing sunspot 5395 rotated into view, revealing a sunspot that was over 40,000 km across (more that fifty times the size of the Earth), covering some 3,600 millionths of the solar disk, with a complex array of umbra contained within a single penumbra, a dramatic example of a classic δ-spot (see Chapter 5). By all accounts, this was an extremely large and magnetically complex active region (Figure 10.3), with all the conditions available to produce many large solar storms.

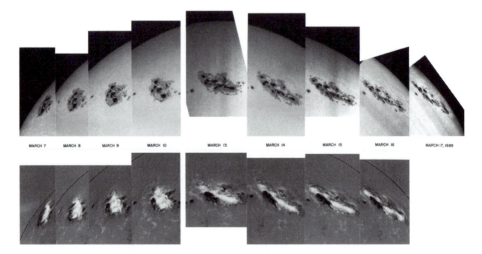

Figure 10.3 A series of images of the Sun taken during the period March 7–17, 1989. This series shows the evolution of a sunspot group as it moves around the Sun. The upper sequence, taken in white light, shows the more usual representation of the solar surface. Below, the same sunspot region is followed from magnetogram studies, where black and white regions show different magnetic polarity. The correspondence between morphology and magnetic activity is clear. (Credit: NOAO/AURA/NSF/WIYN)

Over the period March 6 to March 19, active region 5395 produced almost 200 solar flares, with almost sixty of them being of M- or X-class as described in the preceding chapter. Thirty-nine of these large flares had an associated CME. One of the most powerful solar flares ever recorded occurred on March 6, and this was accompanied by two other very large flares. The GOES satellite X-ray detectors reached X15 (see flare classification in Chapter 9), while the proton detectors saturated and were overwhelmed for around twenty-seven minutes, and radio signals reached 2,000 times the background level. (While this was the largest flare on record at that time, it is only sixth on the all-time list of largest flares—the highest being an X28 flare that occurred on November 4, 2003.) The three flares resulted in a continuous stream of high-energy radiation that was about twenty times more intense than normal and impinged on the Earth for over ten hours. While this intense activity was unprecedented, there was relatively little impact on the Earth until another large event, GOES class X4, occurred on March 9. This time a powerful coronal mass ejection, with a speed in excess of 1.5 million kilometers an hour, was produced with a trajectory directly towards the Earth. Late on March 12, the CME itself reached the Earth. The Earth was pummeled by a blast of high-energy protons that peaked at over one hundred times normal numbers. The effects at the Earth were felt almost immediately. The rapid influx of charged particles generated rapidly varying electrical currents in the Earth's ionosphere, causing an array of effects from a change in the auroral electrojet, a high-altitude electrical current system circulating the Earth at high latitudes, to the generation of induced electrical currents at the ground. The strength of the storm was such that the aurora (northern lights) were seen as far south as Florida and Cuba, with reports from all over the southern United States, including Texas and Mississippi. The induced ground currents caused bigger problems. At the nuclear power plant in Salem, New Jersey, a $10 million dollar transformer was effectively destroyed when currents exceeded safety norms. More dramatically, in the province of Quebec in Canada, an entire electrical grid collapsed, plunging more than 6 million people into darkness and, worse, turning off heating systems in the $-7°C$ temperature. What surprised the utility operators was the sheer speed at which the complete shutdown occurred. Within a space of ninety seconds, a chain reaction had shut down the electrical power to the province including the entire city of Montreal. The blackout resulted in a loss of almost 20,000 megaWatts of power at a total cost to the utility and customers stretching into the hundreds of millions of dollars. In addition, the loss of over 1,000 MW exported to the United States increased the strain on grids to the northeastern United States, which officials have stated almost resulted in a catastrophic shutdown.

In addition to the induced electrical effects of the March 1989 storm, a host of other consequences ensued, including a compression of the Earth's magnetosphere from over 54,000 km to less than 30,000 km, well inside the geosynchronous orbit region where many of the world's telecommunications, weather, and military satellites reside. As a result, many satellites lost their natural protection from the rages of the Sun. On March 13, NOAA's GOES-7 weather satellite suffered circuit problems that rendered it effectively useless for large periods of time, and had such severe radiation damage to the cells of its solar panels that it lost 50% of its power generation. Many other satellites exhibited an array of electrical faults known as single event upsets (SEU), which can temporarily shut down onboard computers without necessarily causing permanent damage. The increased "radio noise" of the Earth's ionosphere disrupted communications both on the Earth and between satellites and the Earth, particularly affecting the Global Positioning System (GPS), which is crucial for navigation and, more frighteningly, the accurate launching of military missiles. Finally, the enhanced heating of the Earth's atmosphere by the combination of the influx of energetic particles and the compression of the magnetosphere increased the density of the upper atmosphere by a factor of five to ten. As a consequence, a stronger drag was felt by satellites orbiting in low-Earth orbit, causing many of the orbits to decay and the spacecraft to drop to lower orbits—the U.S. Air Force Space Command lost track of more than 1,300 orbiting objects that had "fallen" to a lower altitude.

It has been almost twenty years since that large geomagnetic storm, and in that time our use of and reliance on space-based technology has grown exponentially, with huge networks of satellites servicing pagers, cell phones, BlackBerrys, TV, and radio, and in-car and hand-held GPS systems. In addition, electrical transmission networks span longer and longer distances, making more people more vulnerable to a chain effect like that felt in

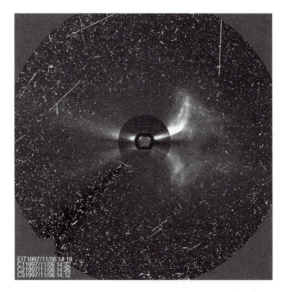

Figure 10.4 Solar storm particles from a large CME hitting SOHO LASCO detector. Courtesy of NASA.

Quebec. The effects of the storm of 1989 provide a warning to us about the possible impacts of solar activity on our everyday lives here on Earth. Continual monitoring of the Sun, better understanding of the physics of solar activity, and improved data and models predicting the likelihood of a solar flare and/or CME are necessary to mitigate the potentially hazardous effects of another storm like the one of 1989.

EFFECTS OF SPACE WEATHER

Space weather phenomena have a major impact for society at large. Our modern reliance on a wide range of technology, from electrical power, and the oil supply for that power, to satellite telecommunications and weather monitoring, makes the twenty-first-century human more susceptible to the whims of the Sun than ever before. The enhanced radiation that signifies a space weather storm can wreak havoc on electronic, electrical, and magnetic systems, with the potential to severely damage our power and space infra-structure. Space weather impacts include:

- Sensitive electronics on satellites, which we rely on for global communications, global security monitoring, and advanced warnings for severe weather, can be temporarily or permanently damaged making the satellites effectively useless.
- The Earth's atmosphere responds to the enhanced energy input by expanding, and as a result increasing, the drag on low-Earth orbiting spacecraft, degrading their orbits and shortening their operational lifetimes.
- Increased radiation damage to solar panels and scientific detectors on spacecraft, reducing the lifetime and scientific productivity of these missions.
- Induced currents in the Earth's ionosphere can cause "short-circuits" in power-grid transformers, resulting in major overload of electrical grids and vast blackouts.
- These same induced currents can accelerate erosion of above-ground oil pipelines, such as those in Alaska, increasing the need for maintenance and repair.
- The larger amounts of radiation produce an increased health hazard for astronauts, and high-altitude polar flight aircraft crews and passengers.
- The induced magnetic activity can affect navigational systems, both human and animal, causing errors in GPS accuracy, which has major implications for fishing fleets and freight ships, while also disrupting the natural navigational systems of animals, such as whales, that rely on magnetic information to steer their way around the world.

Since the launch of the first Earth-orbiting spacecraft, *Sputnik*, in October 1957, more than 4,500 satellites have been launched. Today there are more than 850 satellites in operation, with a further 1,700 or so non-operational. The jobs of the active satellites include telecommunications (e.g., cell phones, satellite radio and TV), environmental monitoring (e.g., the GOES weather satellites), and scientific research. A large number of satellites are used to provide data for models of global weather forecasting, vital in predicting where and when natural disasters, like hurricanes, floods, tidal waves, and forest fires, may occur. Weather satellites also provide important information for agriculture, helping decide best times for planting and harvesting. Satellites, like the Landsat fleet, allow us to monitor the Earth for important studies of land use, long- and short-term animal and human migrations, and pollution monitoring. If you have ever used the satellite image option in *Google Maps* (http://maps.google.com), you will be aware of the utility and power of these space resources. In addition, military reconnaissance satellites help monitor the worldwide security of various nations around the world, while scientific satellites provide new discoveries about the Earth, the solar system, and the universe. Finally, the world today relies heavily on telecommunications satellites for cell phone use, long-distance communications, and a large variety of entertainment options. Even cars today come equipped with satellite radio, satellite-reliant navigation (GPS), and telephones.

The satellites are distributed around the Earth in a variety of orbits, depending upon their function. Satellites in polar or near-polar orbit at high inclinations relative to the Earth's equator travel in a circular orbit over the north and south poles, and can effectively monitor the whole Earth as it rotates beneath them. In contrast, a satellite in a geosynchronous orbit remains stationary above a particular spot on the Earth at a height of almost 36,000 km. At this height the orbital period of the satellite matches that of the Earth. Typically, weather, TV, and communications satellites use these orbits. Finally, many satellites lie in low-Earth orbits at about a height of 320 km, for example, Mir space station, the Hubble Space telescope and the International Space Station all lie in a low-Earth orbit. Figure 10.5 shows a representation of the distribution of Earth-orbiting satellites: the high density of low-Earth spacecraft is clearly evident, as are the rings of geosynchronous and polar orbits. All of these resources, mounting to billions or even trillions of dollars of investment and more importantly providing invaluable life-saving services, are at the mercy of space weather.

THE EFFECTS OF SOLAR RADIATION

High-energy solar radiation comes in two types: photons and particles. The photons represent the electromagnetic radiation with the higher energies, X-ray and gamma-ray ranges, being the more relevant to space weather. The particle radiation is in the form of fast-moving particles, dominated by electrons, protons, and helium nuclei. These particles are accelerated directly by the flare, or result from the shock produced by interaction of a fast CME with the solar wind (Chapter 9). The energy associated with this

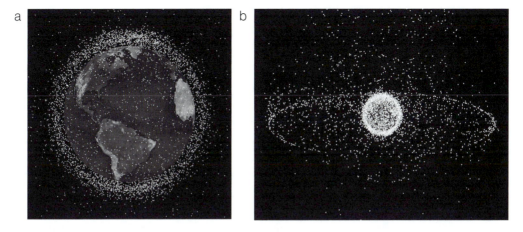

Figure 10.5 Distribution of satellites in (a) low Earth orbit (LEO) and (b) geosynchronous Earth orbit (GEO). The GEO images are generated from a distant oblique vantage point to provide a good view of the object population in the geosynchronous region (around 35,785 km altitude). Note the larger population of objects over the northern hemisphere is due mostly to Russian objects in high-inclination, high-eccentricity orbits. (Credit: NOAO/AURA/NSF/WIYN)

high-energy solar radiation can be transferred to matter, and if that matter is in the form of a human being, then serious health problems can ensue. On the Earth we are protected from this intense solar radiation by the atmosphere and magnetosphere, which stop the photons and particles, respectively; but in space astronauts are subject to potentially lethal dosages of radiation during the largest solar storms. High radiation doses, which can result from being exposed to solar radiation, can damage chromosomes, kill cells, and potentially cause cancer. Severe doses may even cause death. Solar protons with energies greater than 30 MeV are particularly hazardous, and several storms recorded since the early 1970s have generated large numbers of such protons, enough to have produced a lethal dose of radiation had an astronaut been exposed. Fortunately, these storms are relatively infrequent and short-lived such that, to date, no astronaut has been subjected to dangerous levels of radiation. Astronauts are protected against background solar radiation by the metal structure of the shuttle or the space station. However, an astronaut on a spacewalk, or on the fully exposed surface of the Moon, runs a significant risk since these high-energy particles would pass straight through the fabric of a spacesuit. High-energy solar proton storms are also a concern for crew and passengers on board high-altitude air flights, as elevated radiation levels have been detected during enhanced solar activity.

The primary means by which the high-energy radiation (photons or particles) damage a living cell is by ionizing the atoms in the molecules that make up the cell—the incoming radiation has enough energy to effectively "kick" electrons off the atoms in the molecules. The ionized atoms (ions) prevent the cell from functioning properly since the charged atom now interferes with the normal bioelectrical signals that trigger the various cell functions. In the most extreme cases, the DNA in the cell can be damaged. There are two major ways in which the DNA can be affected. One is via the production of free radicals, which are generated when the water in your body absorbs the radiation and subsequently becomes ionized. The free radicals react readily with the DNA molecule, breaking some of the bonds and effectively altering slightly the genetic makeup of the cell. The other is via the direct interaction between the incoming radiation and the DNA molecules that ionize some of the atoms making up the DNA molecule with the same potentially hazardous effects.

Humans are subject to radiation all the time from galactic cosmic rays and naturally occurring radiation sources, such as radon gas. The levels of this radiation are small enough that our health is at no immediate risk. However, the various governments around the world have devised limits for radiation exposure to protect their populations. Radiation dose is measured in units called *rems* (Roentgen Equivalent Man) or Sieverts. A rem is a unit that relates the dose of any radiation to the biological effect of that dose, and one Sievert is approximately 100 rems. The daily dose of radiation for a human being on the Earth from natural sources is about 10^{-3} rems per day (the dose from a standard chest X-ray is about 2 millirems, or about two days of natural radiation dosage). The recommended dosages for the general population and for astronauts are shown in Table 10.1.

The limits for astronauts are set by the National Council on Radiation Protection and Measurements (NCRP). The cells listed in the table are those that reproduce the quickest and are therefore the most liable to damage by radiation. Possible symptoms of high doses of radiation, radiation sickness, include severe burns that are slow to heal, sterilization, organ damage, and potentially cancer and other damage to organs. High doses are rapidly (within days or weeks) fatal.

Solar radiation is a potentially serious hazard for current and future astronauts as we think about colonizing the Moon and Mars. Understanding the

Table 10.1. **Maximum recommended radiation dosage, in rems, for astronauts in comparison with that for radiation workers in nuclear power plants**

CELL	TIME PERIOD	ASTRONAUTS	RADIATION WORKERS
Blood-forming organs	30 days	25	–
	1 year	50	5
Eyes	30 days	100	–
	1 year	200	15
Skin	30 days	150	–
	1 year	300	50

Note: As a reference, the maximum recommended dosage for non-radiation workers is 0.1 rems per year for whole body exposure circumstances on the Sun that lead up to production of enhanced levels of high-energy solar radiation is a crucial component of space weather research. The next section discusses the current status of our capability to predict solar flares, coronal mass ejections, and their effect at the Earth.

PREDICTING SPACE WEATHER

Given the obvious dangers posed by solar flares and coronal mass ejections and their associated high levels of radiation and accelerated ions, there is a real and serious interest in being able to predict when these events will occur and what impact they will have on the Earth. However, like the terrestrial version, predicting or forecasting space weather is not that easy.

The fundamental problem of predicting energetic solar transient events and their consequences does have some parallels to generic prediction problems being tackled in more familiar settings, for example, terrestrial weather, earthquakes, and the stock market. All of these physical systems are complex systems that exhibit both slow evolution and catastrophic change. The events to be predicted have magnitudes that range from very small to very large, and have consequences on local and global scales. What this all means is that even if one gets the physics correct so we know when an event is likely to occur, there is another broader set of physical problems that need to be solved in order to work out the consequences at the Earth, 150 million kilometers away.

The basic problem of predicting the arrival of high-energy solar particles at the Earth, or for that matter anywhere in the solar system, the Moon or Mars for example, can be broken down into three main components, each of which can be treated independently:

1. Prediction of the occurrence of the flare or CME;
2. Prediction of the flux and energy of the energetic particles produced; and
3. Prediction of the path these particles take after being produced.

In addition, one needs to model the modification to the particles on their journey from production to the Earth. The first problem, flare/CME forecasting, requires knowledge of the spatial and temporal patterns at the Sun that lead to an eruption. While in principle this sounds relatively easy, the source regions of these events are constantly changing, with or without producing a flare or CME, and selecting the correct, or even most likely, sequence of behavior from all of the possibilities poses a major challenge. Considerations include the magnetic complexity of the region, the rate at which magnetic flux enters or leaves the region through the photosphere, the effect of surface motions on the magnetic field and how this varies in time, the buildup of electrical currents that carry the free energy required for an eruption to occur, and the development of high-energy latent structures such as prominences and sigmoids.

The recent history of the flaring record for any given active region can also provide information about the likelihood of flaring within that particular region. There is a well-defined pattern of occurrence in flare-producing active regions, with an expectation for the size of a given event based on those that preceded it. This leads to an expectation, or probability that a flare will occur within a given time period, with the accuracy decreasing the longer ahead you predict. If a weather person tells you there is an 80% chance of rain today, you may cancel your pickup softball game. If the same person says that there is an 80% of rain ten days from now, you are unlikely to change your plans, waiting for more accurate information.

Steps 2 and 3 are dependent upon the success of Step 1. However, the ability to predict exactly what a flare and or CME will do after it is initiated requires a complicated and imprecise coupling of physical assumptions, modeling, and data analysis. For instance, flare particles can only affect the Earth (or Moon or Mars) if they are magnetically connected to the flare site, along the Parker spiral. In order to predict the effects of such an event well in advance, one must also predict the speed of the solar wind to determine the field geometry. Even if one is directly connected to the source of the energetic particles, predicting the number and energy of the particles produced is beyond the scope of our current understanding. It is even more difficult when dealing with coronal mass ejections. Fast CMEs produce shocks across a wide swathe of solar longitude, and these shocks accelerate ions up to the large energies observed. The large extent of the shocks makes connectivity of the Earth to the source of high-energy particles relatively likely. However, in order to understand the properties of the shock that forms, shock strength, location, angle to magnetic field, and so forth, one must not only predict that a CME will occur, but how fast it will accelerate, what speed it will attain, and how big it will be. The CME forms a shock by running into the solar wind, so, once again, the expected speed of the solar wind through which the CME will pass will also have to be predicted. Getting this far would be a major achievement, but we still have the problem of predicting the energy and number distributions of the particle accelerated by the shock and how these distributions vary along different Parker spiral field lines.

All of this is further complicated by the fact that unlike the weather forecasters on the Earth, we can only monitor the Sun from distinct locations, mostly on the Earth, in Earth orbit, or at the L1 Lagrangian point, 1.5 million kilometers "upstream" (the *SOHO* and *ACE* spacecraft are located there), which for the fastest and most geo-effective CMEs gives about thirty minutes warning. We have very little direct information from the vast bulk of space between the Sun and the Earth that can help warn of impending solar storms.

Despite the severe pessimism of the foregoing discussion, all is not lost. While we might not be able to model all of the physics required to accurately predict the details of the energetic particles from an observation of a solar active region, scientists can estimate the likelihood that a geo-effective event will occur. The Space Weather Prediction Center (SWPC) based in Boulder, Colorado, is the branch of the National Weather Service that deals with the potential space hazards resulting from solar storms. Officially, the SWPC continually monitors and forecasts Earth's space environment in order to provide accurate, reliable, and useful solar-terrestrial information relating potential space hazards, for use by the commercial, government, and military interests with resources in space.

Forecasters at the SWPC use up-to-the-minute solar data from space-based and ground-based observatories around the world to provide warnings and alerts of impending solar and geomagnetic activity. There are four types of space weather alert messages:

- **Watch** messages, for Geomagnetic A-indices, are issued for long-lead geomagnetic activity predictions.
- **Warning** messages are issued when some condition is expected, and contain a warning period and other information of interest.
- **Alert** messages are issued when an event threshold is crossed, and contain information that is available at the time of issue.
- **Summary** messages are issued after the event ends, and contain additional information available at the time of issue.

A sample alert reads as follows:

Sample Geomagnetic K-index Warning and Alert
WARNING: Geomagnetic K-index of (thresholds 4, 5, 6, or "7 or greater")
expected
Space Weather Message Code: WARK05
Serial Number: 138
Issue Time: 2001 Sep 29 2221 UTC
WARNING: Geomagnetic K-index of 5 expected
Valid From: 2001 Sep 29 2225 UTC
Valid To: 2001 Sep 30 1500 UTC
Warning Condition: Onset
NOAA Scale: G1 - Minor
Space Weather Message Code: WARK07
Serial Number: 140
Issue Time: 2001 Sep 30 0230 UTC
WARNING: Geomagnetic K-index of 7 or greater expected
Valid From: 2001 Sep 30 0230 UTC
Valid To: 2001 Sep 30 1500 UTC
Warning Condition: Onset
NOAA Scale: G3 or greater - Strong to Extreme

The above alert indicated that a strong to severe geomagnetic storm was likely sometime between 02:30 and 15:00 on September 30, 2001, with a K-index of 7 or greater. The warnings and alerts are based on prevailing solar conditions and range of factors that include the rapid increase in X-ray emission indicating a solar flare occurring, enhancements in radio emission, especially those indicating the presence of shocks, rising geomagnetic indices, and growth in numbers and energies of energetic particles.

Space weather is a growing area of scientific interest and public concern. As our dependence on space-based technology grows, we are increasingly susceptible to the vagaries of the Sun. Technology aside, as the human exploration of space continues, the dangers of space weather become more important as increased exposure to solar radiation can have severe health impacts for astronauts.

On January 14, 2004, President George W. Bush announced a new vision for the nation's space exploration program. The president committed the United States to a long-term human and robotic program to explore the solar system, starting with a return to the Moon that will ultimately enable future exploration of Mars and other destinations.

The president's plan for steady human and robotic space exploration is based on the following goals:

- First, the United States will complete its work on the International Space Station by 2010, fulfilling our commitment to our fifteen partner countries. The United States will launch a refocused research effort on board the International Space Station to better understand and overcome the effects of human space flight on astronaut health, increasing the safety of future space missions.
  - To accomplish this goal, NASA will return the space shuttle to flight consistent with safety concerns and the recommendations of the Columbia Accident Investigation Board. The shuttle's chief purpose over the next several years will be to help finish assembly of the station, and the shuttle will be retired by the end of this decade, after nearly thirty years of service.
- Second, the United States will begin developing a new manned exploration vehicle to explore beyond our orbit to other worlds—the first of its kind since the *Apollo Command Module*. The new spacecraft, the *Crew Exploration Vehicle*, will be developed and tested by 2008, and will conduct its first manned mission no later than 2014. The *Crew Exploration Vehicle* will also be capable of transporting astronauts and scientists to the International Space Station after the shuttle is retired.
- Third, the United States will return to the Moon as early as 2015 and no later than 2020, and use it as a stepping stone for more ambitious missions. A series of robotic missions to the Moon, similar to the *Spirit Rover* that is sending remarkable images back to Earth from Mars, will explore the lunar surface to research and prepare for future human exploration. Using the *Crew Exploration Vehicle*, humans will conduct extended lunar missions as early as 2015, with the goal of living and working there for increasingly extended periods.
  - The extended human presence on the Moon will enable astronauts to develop new technologies and harness the Moon's abundant resources to allow manned exploration of more challenging environments. An extended human presence on the Moon could reduce the costs of further

exploration, since lunar-based spacecraft could escape the Moon's lower gravity using less energy at less cost than Earth-based vehicles. The experience and knowledge gained on the Moon will serve as a foundation for human missions beyond the Moon, beginning with Mars.

- NASA will increase the use of robotic exploration to maximize our understanding of the solar system and pave the way for more ambitious manned missions. Probes, landers, and similar unmanned vehicles will serve as trailblazers and send vast amounts of knowledge back to scientists on Earth.

Radiation exposure from space weather becomes increasingly important as more humans venture into space and spend more time there, and human exploration of the solar system takes humans outside the protective envelope of the Earth's magnetosphere. The Moon has no atmosphere or magnetic field to shield it from the direct exposure to solar radiation. The trip to Mars takes eighteen months, where the astronauts will rely on the protection of their spacecraft to protect them. A working knowledge of space weather becomes crucial to the success of these missions. As observations of the Sun improve, as physical models of solar phenomena become more realistic, and as prediction tools become more sophisticated, our understanding of space weather and the physics that drive it will ultimately increase our capability to warn satellite providers, power station engineers, and astronauts on their way to Mars that a solar storm is on its way.

RECOMMENDED READING

Carlowicz, Michael, and Ramon Lopez. *Storms from the Sun: The Emerging Science of Space Weather.* Washington, DC: National Academies Press, 2002.

WEB SITES

Exploration science: http://www.nasa.gov/mission_pages/exploration/main.
National Council on Radiation Protection and Measurements: http://www.ncrponline.org.
National Space Biomedical Research Institute: http://www.nsbri.org.
National Space Weather program: http://www.nswp.gov.
NOAA geomagnetic storms: http://www.swpc.noaa.gov/NOAAscales/#Geomagnetic Storms.
Space Weather Now: http://www.spaceweather.com.
Space Weather Prediction Center: http://www.swpc.noaa.gov.

11

The Sun and Climate

If not for the presence of the Sun, the Earth would be a cold, lifeless lump of rock floating in the void of space. The Sun generates sufficient energy that even at a distance of 150 million kilometers, the surface of planet Earth is heated to around $-18°C$. This may seem rather cold but the presence of an atmosphere containing infrared absorbing greenhouse gases prevents a lot of the heat from reradiating back into space, raising the surface temperature of the planet by a further 30–35 degrees. The actual surface temperature of the Earth, averaged over the whole planet, is about $15°C$. The greenhouse gases, mostly water vapor and carbon dioxide, are provided by volcanic outgassing and serve to warm the planet by trapping some of the infrared radiation emitted by the Sun-heated surface of the planet. The presence of the Earth's atmosphere increases the temperature sufficiently that water is not frozen, but liquid, and this is a crucial component in the formation of life on Earth.

Life on Earth depends on many factors, not least of which is the stability of the climate over long time scales. The presence of a stable climate relies on both internal and external conditions that govern both the input of energy to the Earth, and its variations, and the response of the planet to this variability. The geological, chemical, and biological components of the Earth all work together in response to the external heat and light from the Sun to create the delicate balance required to maintain a stable climate over millions of years. Geological activity creates an atmosphere that warms the planet, generating conditions for liquid water to exist, providing an environment in which life can develop, producing oxygen that makes its way into the atmosphere, where it produces ozone that protects the surface from

173

Temperature of a Planet

The temperature of a planet is essentially governed by its distance from the Sun, which determines the amount of heat available to be absorbed, and its reflectivity or *albedo*, which determines how much of the available heat is actually absorbed. To calculate the average temperature of the planet, we need only find a balance between the energy absorbed and that reradiated as heat. On the left-hand side of this balance is how much energy is available from the Sun at a distance of the planet, say the Earth. This is known as the solar constant, defined in Chapter 2, and will be discussed in more detail later in this chapter. The solar constant is 1,366 Watts per square meter. Thus, every square meter at the distance of the Earth receives 1,366 Joules of energy every second (1 Watt = 1 Joule/s). On the right-hand side of this balance is how much of this energy is absorbed by the Earth. The total absorbed is the amount available minus the fraction that is reflected, known as the albedo. The amount of radiation reflected is dominated by the reflectivity of clouds and the ice caps. The average reflectivity of the Earth is approximately 30%, meaning that 70% of the sunlight available for heating is absorbed. Thus, the amount of solar energy absorbed by the Earth in one second is $0.7 \times 1{,}366$ Watts per square meter $= 956.2$ Wm^{-2}. The area of the Earth lit up by the Sun is the circular area $A_c = \pi R_e^2$ and so the total amount of energy absorbed per second is $E_{in} = 956.2 \times \pi R_e^2$ Watts.

To be in balance with this energy intake, the Earth has to radiate an equivalent amount of energy across the whole surface of the Earth. The area of the sphere is $4\pi R_e^2$ and the heat radiated is given by a physical law known as the Stefan-Boltzmann Law, which states that the total radiation is given by a constant $\times$ the surface area $\times$ the fourth power of the temperature, that is, $E_{out} = \sigma \times 4\pi R_e^2 \times T^4$ Watts, where $\sigma = 5.7 \times 10^{-8}$ $Wm^{-2}K^{-4}$. Equating E_{out} with E_{in} yields the temperature of the Earth that would result purely from heating by the Sun: $\sigma \times 4\pi R_e^2 \times T^4 = 956.2 \times \pi R_e^2$. We then have $T^4 = 956.2/4\sigma$ yielding $T = 254$ K $= -18°C$.

harmful solar radiation, allowing more complex life to develop, and so on. The energy input that drives the system is, of course, provided by the Sun and modulated by the Earth's orbit around the Sun and the tilt of the Earth's axis. Any changes in this system, such as a period of excessive volcanic activity, result in a chain reaction of effects, which over time serve to bring the Earth back to balance. This natural balance is therefore sensitive to permanent changes in either the internal or external conditions. As we near the end of the first decade of the twenty-first century, there is a major concern that just such a permanent change may be occurring. The overwhelming scientific evidence is that the Earth is warming at an alarming rate and that the bulk of this warming is a direct result of human activity—permanently changing the landscape by deforestation for farming, burning fossil fuels in cities, and increased populations of livestock, all of which contribute to the increase of the concentration of greenhouse gases in the atmosphere.

As discussed in earlier chapters, the Sun, like other similar stars, exhibits variability across a wide range of timescales and magnitudes. The eleven-year solar cycle (Chapter 6) is the most evident sign of a regular change in the Sun's brightness. However, the total change in the Sun's luminosity over the course of a solar cycle is less than one tenth of one percent, and this is so small that the total radiation output of the Sun is known as the solar constant. This small level of variability is not thought to have a significant impact on the Earth's climate. Modern space-age measurements of the Sun's radiation across the whole electromagnetic spectrum have shown that the solar constant actually varies, with the variations showing up on timescales as disparate as hours to decades in addition to the obvious eleven-year cycle.

To understand changes in the Earth's climate and to assess its cause, both current and historical trends, we first need to understand the natural variations and how these relate to variations in the Sun, the source of the energy driving the Earth's climate. This chapter takes a look at the impact of solar variability on the Earth's climate and what the results imply for the debate on global warming.

Climate versus Weather

In arguments about climate change, there is often confusion regarding the difference between weather and climate. Most people are familiar with, and frequently unimpressed by, the accuracy of weather forecasts that describe the likely change in local conditions over a short space of time. Weather can change a lot in a very short time. On the other hand, climate describes the average weather conditions and regular cycles in the pattern of weather over a period of years (often decades or centuries) in a given region. Scotland is generally wetter and cooler than New Mexico, (climate) even though on any given day parts of Scotland may be hotter and drier than parts of New Mexico (weather). In other words, climate describes the typical state of the weather for a particular region at a particular time of year.

Climate can also be used in a broader sense to describe the large-scale state of a habitable environment such as the Earth. A description of a planetary climate incorporates the state of the atmosphere, the oceans, and the land masses, as well as the contributions from ice covering, which affect the radiation balance of the Earth, and life, which affects the compositions of the air and water.

VARIATIONS IN THE EARTH'S CLIMATE

While the Sun's energy input to the Earth is a crucial factor in determining climate and climate changes, the amount of solar heating received by the Earth can vary as a result of factors other than intrinsic changes in solar brightness. For example, long-term changes in the Earth's orbit alter the amount and distribution of sunlight reaching the Earth, and this is thought to be the primary cause of the major ice ages evident in the Earth's climate records.

Throughout its history, the Earth has been subjected to numerous ice ages where a large fraction of its surface was covered with snow and ice. Ice ages occur when the average temperature of the planet as a whole falls by three or four degrees for a sustained period. There have been at least four major ice ages over the history of the Earth, with the most recent ending only 10,000 years ago. While the causes of the ice ages are not known exactly and can involve a complex sequence of events, several theories exist, each probably containing some aspects of the whole picture. Changes in solar brightness over hundreds of millions of years influence the development and recession of ice ages while the continually changing continental plate configurations also impact the formation and cessation of ice ages over similar time scales. The coming and going of ice on shorter time scales, say millions of years, is thought to be influenced by the cyclical changes in the Earth's orbit, commonly known as the Milankovitch cycles after the Serbian scientist, Milutin Milankovitch (1879-1958), who first put forward the idea. Gradual changes in the shape and orientation of the Earth's orbit about the Sun and in the tilt angle of the Earth's rotation axis occur as a result of the long-term effects of the small gravitational pull of the other planets in the solar system, most notably Jupiter, and the Moon. The combination of these gravitational effects introduces cyclical changes in the Earth's orbit with periods of order 19, 23, 41, and 100 thousand years, with consequent periodic changes in the amount and distribution of sunlight hitting the Earth. The seasons of the Earth are determined by the tilt of the Earth's axis relative to the ecliptic plane. This tilt varies in two ways. First, the location in the sky that the axis points varies continuously, technically the axis of the Earth's rotation precesses, with a period of 26,000 years. Presently, the Earth's rotation axis points to the North Star, Polaris, but 13,000 years from now it will point to the star Vega. Second, the magnitude of the tilt also changes with time, ranging from 22 degrees, through its present value of 23.5 degrees, to about 25 degrees. The changing tilt affects the strengths and the durations of the seasons and, consequently, can help foster conditions for an ice age to occur—smaller tilt means longer polar winters and longer lasting ice. Ice ages have been found to correlate well with periods of smaller-than-average axis tilt.

Glacials and Interglacials

During the last 2 million years of Earth's history, a period known as the Quaternary Period in geology, there have been several periods where polar ice sheets expanded significantly equatorward. The Quaternary Period is divided into two distinct epochs: the Pleistocene and the Holocene. We currently reside in the Holocene period, which extends back only the last 10,000 years. The periods of extended ice sheets are known as *glacials*, and the intervening times when the ice sheets have retreated are known as *interglacials*. Glacial periods are characterized by cooler average global temperatures and drier average climate. Mountain glaciers extend to lower altitudes, sea levels drop, and oceanic circulation patterns may be disrupted. During interglacial periods, the temperatures rise, the ice sheets contract, and the climate becomes similar to what we have at the present time—the present Holocene epoch is considered to be an interglacial.

Our knowledge of the climatic behavior over the last 2 million years comes from the measurement of two distinct isotopes of oxygen found in the shells of dead plankton. During glacial periods, the lower sea levels preferentially result in the reduction of one of the isotopes. The relative amounts of these isotopes allow us to determine periods of cooler-than-average temperatures. Scientists estimate that the last glacial period began some 120,000 years ago and ended about 10,000 years ago.

THE GREENHOUSE EFFECT

One of the most important facets for the presence and sustainability of life on Earth is the greenhouse effect. The greenhouse effect describes the excess warming of the Earth as a result of the presence of particular gases in the Earth's atmosphere. These gases act as a thermal blanket, preventing the escape of solar energy absorbed by the Earth's surface. Heat and light from the Sun impinges on the Earth. Some of this radiation is reflected back into space by clouds, ice, and water, and the rest is absorbed and heats the surface. The heated surface then radiates infrared radiation back out to space through the atmosphere. Greenhouse gases in the atmosphere (water vapor, carbon dioxide, methane, and ozone are common examples) are efficient absorbers of infrared radiation and prevent a significant fraction of the radiated heat from escaping the Earth, resulting in a warming effect (Figure 11.1). Without these gases, the Earth's temperature would be almost 40°C cooler and too cold for liquid water to exist in any great amounts (see sidebar earlier in this chapter). So, greenhouse gases are an important component of the climate control of the Earth. Without them life could not exist. However, significant changes in the amounts of atmospheric greenhouse gases can have major consequences for the Earth's climate.

The main greenhouse gases are water vapor and carbon dioxide (CO_2). Water vapor is thought to be responsible for 40%–70%, with CO_2 a further 9%–25%. Overall, the most abundant and dominant greenhouse gas in the atmosphere is water vapor, although it is not long-lived, nor does it mix well with other atmospheric constituents. The concentration of water vapor in the atmosphere varies from 0% to 2% by location and can exist in several physical states including gaseous, liquid, and solid. The overall average concentration is not believed to be directly affected by human activity. However, the effect of increased amounts of other greenhouse gases can indirectly affect the concentration of atmospheric water vapor—a warmer atmosphere can hold more water, for example.

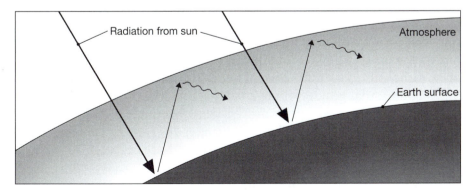

Figure 11.1 The Greenhouse Effect. Illustration by Jeff Dixon.

The levels of CO_2 in the Earth's atmosphere were measured at 382 parts per million in 2006, 36% higher than preindustrial measures, with the indication that most of this increase is due to human activity. Methane, ozone, and man-made chlorofluorocarbons also contribute to the greenhouse effect. Methane results naturally from vegetative decay in swamps, but human activity may now account for over 50% of the total from growing rice in paddies, burning of forest land, and the increase of livestock populations. Current estimates place the concentration of methane in the atmosphere at 1.76 parts per million, much lower than the 382 parts per million recorded for CO_2 in 2006, but represents a 148% increase over preindustrial measures. Methane is also a much more effective greenhouse gas being about 20 times more potent at absorbing infrared radiation, so its contribution to the predicted global warming may be as high as 20%.

A key factor in the global climate conditions of the Earth is the amount of carbon dioxide, a greenhouse gas, in the solar atmosphere. As mentioned earlier, current estimates place the concentration of CO_2 in the Earth's atmosphere at close to 400 parts per million, which represents a 36% increase since preindustrial times. The concern is that human production of CO_2 is rapidly changing the climate by resulting in substantial increases in the average global temperature. Too much CO_2 in the atmosphere and the planet warms up, with potential devastating effects with melting polar ice caps, rising oceans, and marked changes in ocean and wind currents. Too little CO_2 is also a problem as it would lead to an overall cooling of the Earth, possibly below the freezing point of water, which would limit the availability of liquid water and have severe consequences for life on Earth. Fortunately, the Earth has a natural control mechanism to regulate the amount of CO_2 in the atmosphere—the carbon dioxide cycle.

Atmospheric CO_2 comes from a number of natural sources: volcanic outgassing, plant decay, and animal exhalations. Once in the atmosphere, the carbon dioxide dissolves in rainwater to form a slightly acidic mixture. The amount of CO_2 that dissolves is sensitive to the temperature of the atmosphere—the hotter the atmosphere, the more CO_2 dissolves. The acidic rain then erodes the rocks on the surface and mountains of the Earth, the resulting minerals being carried downstream in the rivers and streams to the oceans. A chemical reaction occurs where the calcium in the eroded rock reacts with carbon dioxide dissolved in the water to form carbonate minerals, which fall to the ocean floor. The accumulated deposits eventually form carbonate rocks, such as limestone, which then get carried via the ongoing terrestrial geology of plate tectonics to subduction zones, regions where the dense seafloor meets the less dense continental shelf, forcing its way under the continental rock to the mantle. The heat of the mantle frees the trapped carbon dioxide in the carbonates and releases it once again into the atmosphere via outgassing from volcanic activity.

The CO_2 cycle, then, regulates the Earth's temperature since too much CO_2 increases the temperature, which increases the rate at which chemical reactions with water removes the CO_2, which then results in less greenhouse effect and more cooling. Too little CO_2 results in cooler temperatures slowing the rate of removal of CO_2 from the atmosphere, the levels of which are then replenished via continuing volcanism. At present, on the Earth, there is sixty times as much CO_2 dissolved in the oceans than is present in the atmosphere, and over 170,000 times as much entrapped in rocks.

This sounds good for global warming, since however much excess CO_2 we pump into the atmosphere from human activity the Earth can always reprocess it via the carbon dioxide cycle. The problem for humans is that the temperature regulation process can take as much as 200 million years!

SNOWBALL EARTH

Some of the ice ages have been so strong and so long that the scenario of the *Snowball Earth* has been developed. There is evidence for three ice ages where the ice coverage extended all the way to the equator. The evidence consists of the distribution of glacial deposits around the world and associated carbon isotope anomalies in seawater at given geological times, the discovery of soil deposits associated with glaciation at equatorial latitudes, corrected for geological time, and a host of other more subtle signs. For details see www.snowballearth.org. The Snowball Earth ice ages are thought to have occurred 2.2 billion years ago (Makganyene) lasting millions of years, 710 million years ago (Sturtian) lasting millions of years although more a precise duration is difficult to assess, and 650 million years ago (Marinoan) lasting 6–12 million years. The last Snowball Earth ended about 635 million years ago. The names come from the names of the glacial deposits that indicate the presence of the ice age. Modern computer models suggest that global temperatures during these periods were around $-50°C$.

As the temperature cools and the oceans freeze, the ice sheets grow to cover more and more surface area. This in turn, increases the reflectivity, or albedo, of the Earth, and so more sunlight is reflected and less absorbed, further reducing the temperature and leading to more of the oceans freezing to form ice, which reflects more and so on (Figure 11.2).

Why then doesn't the Earth remain covered in ice for all time? The secret is the carbon dioxide cycle. Volcanic activity is essentially unaffected by the surface temperature and the presence of ice. Thus, ongoing volcanic activity continues to contribute large amounts of CO_2 to the Earth's atmosphere, which cannot be absorbed by the ice. The presence of such large amounts of this greenhouse gas serves to act like a thermal blanket, warming the planet and ultimately causing the ice to melt. The less ice, the less reflectivity, the more sunlight absorbed, and the warmer the planet becomes until the ice recedes back to the polar regions. The excess carbon dioxide is then processed via the carbon cycle and the Earth returns to "normal." This whole process is thought to last on the order of 10 million years.

Figure 11.2 Snowball Earth.

In addition to the external factors that affect the impact of the Sun's heating of the Earth, we must also consider the intrinsic variability of the Sun over long, medium, and short timescales as this has an important bearing on how the present climate was established and maintained.

THE SUN'S INFLUENCE

In 1843, German amateur astronomer Heinrich Schwabe noted that the number of sunspots apparent on the solar disk waxed and waned with a regular eleven-year period, which we now know as the solar cycle (Chapter 6). This regularity prompted many people to report correlations between the Sun's variability and all manner of terrestrial and human events, including weather patterns, good and bad harvests, outbreaks of disease, and even how well the economy was doing. So began the early suggestions of a connection between solar variability and climate on the Earth.

Of specific interest to the debate on the influence of sunspots on the Earth's climate was the period known as the Little Ice Age (see Chapter 6). There is significant disagreement as to when the Little Ice Age started, with some scientists arguing that the advancement of Arctic ice beginning in the year 1250 was the first sign of the changing climate, and others arguing that the climatic temperature minimum of 1650 defines the true start. There is, however, general agreement that the end occurred around 1850, when temperatures began to warm and weather patterns began to stabilize. What is notable from the solar perspective is that the Little Ice Age, and in particular the periods of minimum temperatures, coincided with extended periods of low sunspot activity. Four distinct periods known as the Wolf (1280–1350), Spörer (1450–1550), Maunder (1645–1715), and Dalton (1790–1820) sunspot minima accompanied the coldest times of the Little Ice Age. A recent model by Drew Shindell of NASA has determined that the lack of sunspots during the Maunder minimum, and correspondingly dimmer Sun, reduced the magnitude of the westerly winds in the atmosphere, cooling the continents during winter. The changes were primarily a result of the lower formation of ozone in the stratosphere because of the decreased UV emission from the Sun during this time. There are also studies indicating that the length of the solar cycle correlates well with global temperature changes with shorter cycles, leading to warmer-than-average temperatures.

Radiance versus Irradiance

When discussing the solar forcing of climatic effects on the Earth, we refer to the irradiance of the Earth by the Sun. *Irradiance* is defined as the power received per unit area of a surface from all directions of the overlying hemisphere. However, *radiance* is defined as the power emitted per unit area of the emitting source along a cone with unit solid angle. In other words, *solar radiance* is the amount of electromagnetic energy put out per second in a given direction, while *solar irradiance* is the amount of solar energy hitting the surface of an object such as the Earth.

It is perhaps not surprising that there is a connection between the variation of solar brightness with heating and cooling episodes on the Earth. After all, the Sun provides all of the energy that drives the Earth's vast land, air, and ocean systems, creating weather, predictable or otherwise, and a generally stable climate. However, measuring long-term climate change in association with solar variability requires an understanding of how the Earth responds to heat and light from the Sun. The physics and chemistry of this can be quite complicated, especially when the role played by the Earth's geology and the contribution made by the presence of life are factored in. However, some basic, but powerful, relationships can be made.

One of the key factors is the amount of direct sunlight absorbed by the Earth, scientifically known as the *solar irradiance* of the Earth (see sidebar). The effects of differing amounts of direct sunlight are already well known to us—it is this simple fact that leads to seasons during the year. The tilt of the Earth's axis means that for half of the year the northern hemisphere is tilted towards the Sun, and for the other six months the southern hemisphere. The effect of this tilt is that from late June to late September the northern hemisphere receives more direct sunlight than the southern hemisphere, leading to northern summer, while the southern hemisphere receives less, for southern winter. Changes in the amount of sunlight received clearly can have a marked effect on the weather.

The Sun emits radiation across the whole electromagnetic spectrum, with the largest fraction being confined to optical wavelengths. As discussed in Chapter 6, solar variability shows to different degrees depending upon the wavelength considered. For example, the optical emission varies by less than 0.1% between solar maximum and solar minimum, whereas the X-ray

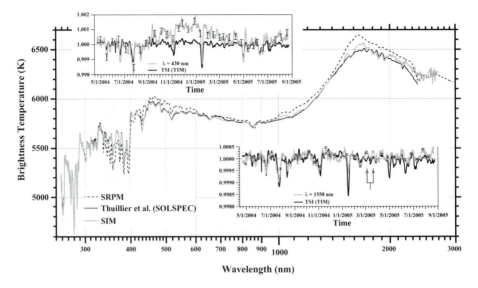

Figure 11.3 The Sun's variations as seen by SORCE's Spectral Irradiance Monitor (SIM). Courtesy of Jerald Harder, Solar Radiation and Climate Experiment, Laboratory for Atmospheric and Space Physics, University of Colorado, private communication.

emission from the solar corona can vary by factors of thirty or more. In addition, the Earth's atmosphere is transparent to some wavelengths, but opaque to others. This all affects the amount of solar radiation that makes it to the Earth's surface (Figure 11.3).

Most of the solar heating of the Earth results from sunlight at visible and near-infrared wavelengths (400–1400 nm); about 70% of the radiation at these wavelengths is absorbed by the surface land and sea. Any variation in the total radiation output of the Sun will drive an associated change in the amount absorbed by the Earth and result in a change in temperature.

The Solar Constant

The bolometric luminosity of the Sun defines the total radiation output emitted regardless of the wavelength of the radiation. The Sun's peak output falls in the optical range (see Chapter 2), but there are significant contributions from UV wavelengths and across the electromagnetic spectrum. As the radiation leaves the Sun, it spreads out across concentric spheres such that the amount of radiation per square meter decreases as the square of the distance from the Sun. At the average distance of the Earth from the Sun (1.496×10^8 km) the solar radiation is measured to be 1.366 kilo-Watts per square meter (1.366×10^3 Wm^{-2}; 1 Watt = 1 Joule per second). The actual value above the Earth varies by 6.9% over the year due to the elliptical orbit of the Earth changing the actual distance from the Sun. Despite the marked variations of the solar cycle (Chapter 6) the distance-corrected value varies by less than 0.1%, and so is referred to as the *solar constant*. This is the amount of radiation that passes through a square meter area above the Earth every second. The amount that actually reaches the surface of the Earth varies dramatically with time of day, cloud cover, latitude, and weather conditions. The maximum amount of radiation to reach the Earth's surface varies between 800 and 1000 Wm^{-2}.

We define the total solar irradiance (TSI) as the total radiant power (energy per second) emitted by the Sun over all wavelengths that impinges on each square meter at the distance of the Earth (see sidebar on solar constant). Attempts to measure the TSI go back to the experiments of American astronomer John Herschel (1792-1871) over 170 years ago, but the limitations of making such observations from the ground, that is having to deal with the variable absorption characteristics of the atmosphere, prevented any definitive measurements from being made. High altitude balloon measurements in the early 1900s could measure the TSI reasonably well, but the instruments were not sensitive enough to be able to detect the 0.1% changes in the TSI with the solar cycle. Accurate measurements of the TSI and its variability have only become possible with the advent of spaceborne instruments that can make accurate long-term measurements at a high enough sensitivity, without the interference of the atmosphere. These observations started in 1978 with the *Nimbus-7* satellite and have continued through a succession of different missions, to the present day. The accumulated results of these missions are shown in Figure 11.4. The plots show variation on several time scales. Most obvious are the twenty-seven-day solar rotation period, the daily variations associated with active regions, and the changes associated with the solar cycle.

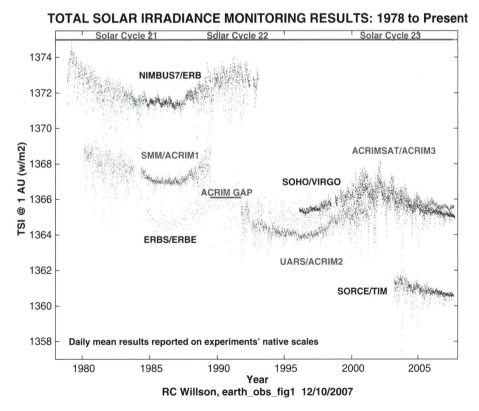

Figure 11.4 Total Solar Irradiance variance from a range of monitoring satellites. Courtesy of NASA.

The variations can largely be explained by the changes in the area coverage of sunspots and faculae. Sunspots tend to reduce the brightness of the Sun, while their associated faculae tend to increase the brightening. The net effect is that the faculae dominate, and the Sun is brighter when there is more sunspot activity. The figure clearly shows that the average TSI varies by about 0.1% over the course of a solar cycle, and models indicate that this corresponds to about a 0.2°C variation in the global average surface temperature of the Earth. However, the oceans are slow to respond to any temperature change, and so the actual temperature response at the Earth is likely to be much lower than this, and therefore negligible from a global warming perspective.

While there are no direct measurements of the TSI for times before spaceborne observations, and routine sunspot observations only go back 400 years, a longer record of solar influence on the Earth's climate can be determined using proxies of the TSI. The direct correlation between the sunspot cycle and the TSI allows scientists to make connections between other indicators of climate such as tree ring records, ice cores, and radioactive isotopes, all of which are affected by variations in solar activity. A careful calibration of these proxies can allow estimates of the solar influence on Earth going back as much as 10,000 years. The evidence from these records is discussed later in this chapter.

The variations in TSI shown in Figure 11.4 represent the integrated change across the electromagnetic spectrum. The variation at specific wavelengths can be significantly more evident, and this is particularly important for shorter wavelengths such as ultraviolet and X-ray radiation. Unlike the optical and near-infrared wavelengths that pass unhindered through the Earth's atmosphere, shorter wavelength UV radiation gets absorbed high up in the atmosphere where it provides a local heating and interacts with oxygen molecules to form ozone. This heating and the associated chemical reactions can have a significant effect on the structure and chemistry in the stratosphere, which in turn can affect the tropospheric climate. The increase in stratospheric ozone associated with higher solar activity warms the upper atmosphere, and this affects the air currents and winds all the way down to the surface. Solar UV radiation varies by as much as 3.5%–7% over the course of a solar cycle, and this can have a marked influence on atmospheric wind patterns and associated climate changes.

GLOBAL WARMING

Global warming is the term given to the measured increase in the average global temperature of the Earth's air and oceans over the last few decades and the expectations for its continuation (Figure 11.5). While the fact that the Earth has warmed significantly over the last three or four hundred years is generally accepted, the causes, natural or anthropogenic, and their relative contributions are hotly debated. In addition, projections for the next

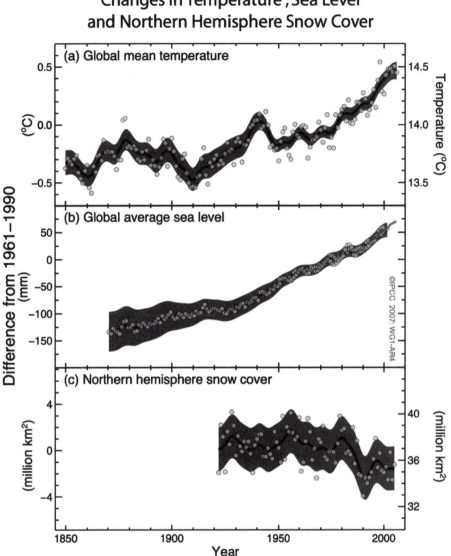

Figure 11.5 Measures of global warming. Growth in the global mean temperature, global average sea level, and decrease in northern hemisphere snow cover. Observed changes in (a) global average surface temperature; (b) global average sea level from tide gauge and satellite data; and (c) northern hemisphere snow cover for March–April. All differences are relative to corresponding averages for the period 1961–1990. Smoothed curves represent decadal averaged values, while circles show yearly values. The shaded areas are the uncertainty intervals estimated from a comprehensive analysis of known uncertainties (a and b) and from the time series (c). WGI FAQ 3.1 Figure 1, Figure 4.2, Figure 5.13, Figure SPM.3 from Intergovernmental Panel on Climate Change Summary for Policymakers.

century, based on various models of the Earth's climate, indicate that the warming trend will continue and could result in a wide range of global effects. The predictions of the climate models, regardless of the differences in the assumptions made, all point to an increasingly warmer planet, with varying degrees of severity of the ensuing problems.

opened for signature in the Rio de Janeiro Summit in 1992 and entered into force in 1994. Subsequent reports were released in 1995, leading to the Kyoto Protocol, in 2001, and, most recently, the fourth report was released in 2007 to a flurry of news reports and media coverage. It provides the overall policy framework for addressing the climate change issue. In December 2007, the IPCC was honored with the Nobel Peace Prize, which it shared with former U.S. Vice President Al Gore "for their efforts to build up and disseminate greater knowledge about man-made climate change and to lay the foundations for the measures that are needed to counteract such change" (http://nobelprize.org/nobel_prizes/peace/laureates/2007/).

In addition to assessing the latest scientific knowledge of global warming to date, the IPCC collates and summarizes the predictions for future global warming and its potential impacts from a wide range of independent climate models. These models indicate that the global surface temperature of the Earth will continue to rise throughout the twenty-first century, with average temperatures rising by 1.1 to 6.4°C above current values. The range in the predicted rise comes about from differing model assumptions about the concentrations of greenhouse gases in the atmosphere and the sensitivity of the climate to this. It is important to point out, however, that all of the models predict increased global warming from present-day values, only disagreeing in the severity of the problem. Furthermore, the slow response of the Earth's oceans to the change in temperature means that warming and sea level rise is expected to continue for at least a thousand years even if greenhouse gas concentrations are stabilized.

The most dramatic changes predicted occur on regional scales, and some of the effects can be summarized as follows (Figure 11.7):

- Warming is greatest over land and at northern latitudes
- Arctic late-summer ice may disappear almost entirely by the latter part of the twenty-first century
- Likely increase in the intensity of tropical cyclone intensity
- Very likely increases in precipitation at high latitudes, with likely decreases in most subtropical land regions.

Report of Intergovernmental Panel on Climate Change

The global warming debate recently took center stage with the publication of the report of the Fourth Intergovernmental Panel on Climate Change (IPCC). The IPCC is a body of more than 1,000 climate scientists across the world set up by the World Meteorological Organization and the United Nations Environment Programme to provide an authoritative statement of scientific consensus on the understanding of global climate change. The IPCC report can be obtained from http://www.ipcc.ch/. The fourth IPCC report raises major concerns for the potential consequences of inaction on the global-warming issue, including worldwide coastal flooding, species extinction, and more severe weather events. The Summary for Policy Makers, a précis of the entire document, underlines the urgent need for action to reduce and reverse decades of human-induced growth of greenhouse gases in the atmosphere.

Key points from the report include the following:

- A 1.5°C change could lead to the risk of extinction of 20%–30% of plant and animal species, while a 3.5°C change could raise this fraction to 40%–70%.
- Oceans could become more acidic, threatening a host of organisms.
- More extreme weather events, spread and duration of droughts, floods and life-threatening storms.
- Decreased water resources for heavily populated centers around the world.

The summary points out that the adoption of more aggressive policies designed to combat climate change can mitigate most of these concerns and be accomplished without damaging national economies, noting that "There is high agreement and much evidence of substantial economic potential for the mitigation of global greenhouse gas emissions over the coming decades" if governments adopt the right policies and incentives.

THE BRIGHTENING SUN

According to our best understanding of how stars evolve, the solar constant is not constant over long times, but has been continually increasing throughout its main sequence lifetime (see Chapter 2). The brightening

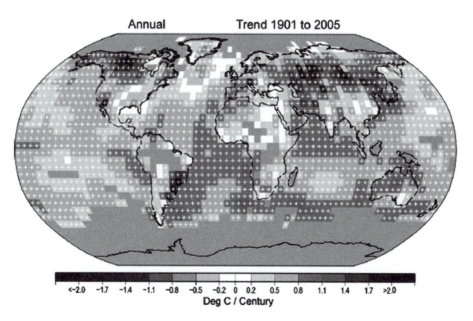

Figure 11.7 Linear trend of annual temperatures for 1901 to 2005 (left; °C per century). Areas in grey have insufficient data to produce reliable trends. The minimum number of years needed to calculate a trend value is sixty-six years for 1901 to 2005 and eighteen years for 1979 to 2005. An annual value is available if there are ten valid monthly temperature anomaly values. The data set used was produced by NCDC from Smith and Reynolds (2005). Trends significant at the 5% level are indicated by white + marks. From Intergovernmental Panel on Climate Change Summary for Policymakers.

Sun is a consequence of the conversion of hydrogen to helium in the nuclear fusion process in the Sun's core. As time progresses, more helium is generated from hydrogen, changing the composition, density, and, ultimately, temperature of the core region. The higher core temperatures that result enhance the rate of the nuclear reactions, which, in turn, generate more photons, increasing the overall brightness of the Sun.

A crude formula for the expected change in the brightness over time was given by English helioseismologist Douglas Gough in 1981 as:

$$L(t) = \frac{L_0}{[1 + 0.4\,(1 - t/t_0)]}$$

where L_0 and t_0 represent the present luminosity and age of the Sun, respectively [$L_0 = 3.86 \times 10^{26}$ W, $t_0 = 4.6 \times 10^9$ years] while $L(t)$ is the luminosity of the Sun at an earlier time t. Thus, one billion years ago, the Sun was only 92% as bright as it is today. This changing brightness would mean a changing solar constant, and therefore a change in the radiation balance of the Earth, and, thus, its climate. Calculations show that the global average temperature of the Earth is very sensitive to the solar luminosity, and consequently the solar constant. One billion years ago, the Earth would have been approximately five degrees cooler, assuming the albedo and concentration of greenhouse gases was essentially unchanged from the present values. However, cooler temperatures would have meant more ice cover, which would mean higher reflectivity (albedo) lowering the average temperature even further.

The Faint Sun Paradox

Models of the greenhouse effect over time in the Earth's atmosphere indicate that lower solar luminosity in the past would have produced an average global temperature of the Earth below the freezing point of water until the Earth was about 2 billion years old. This causes a problem since geological features found on old (>3 billion years) rocks clearly show the signs of flowing water, for example, mud cracks and ripple marks. Sedimentary rocks have also been aged at over 3.8 billion years, suggesting that liquid water was present within the first billion years of the Earth's existence. This contrast between climate models of the early Earth and the geological record is known as the *faint Sun paradox*. Clearly, the assumptions of the climate models are wrong, with the most likely candidate being that the effect of greenhouse gases is underestimated. The carbon dioxide cycle discussed in an earlier sidebar may provide the answer. A lower surface temperature due to a fainter Sun would result in less CO_2 being removed from the atmosphere by rainfall, and consequently the concentrations would increase, limiting the cooling effects of a smaller solar constant.

The increasing luminosity of the Sun over time begs the obvious question: Can't the increasing luminosity of the Sun be responsible for global warming? As discussed earlier, the Earth has warmed by about 1°C over the last hundred years. If this was due to changes in the solar brightness, then such a temperature change would correspond to an increase in the solar constant of about 0.8%. A quick look at the previous luminosity equation would quickly tell you that in such a short time the luminosity of the Sun hardly changes at all (about 0.1 millionths of a percent), and so the long-term brightening of the Sun cannot be responsible for such a marked change in temperature.

However, some other changes in the Sun have been noticed over the last four centuries, since astronomers started observing sunspots (Chapter 5), which require further investigation if we are to fully determine the effect of the Sun on the recent climatic changes. These changes have coincided with the observed warming trend since the seventeenth century, determined from climate reconstructions spanning the last thousand years. This is the period containing the Little Ice Age (Chapter 6). The 350-year warming trend has rapidly accelerated in the last one hundred years, and this rapid growth has been attributed to increases in the atmospheric greenhouse gas concentrations since the industrial revolution. Before such a definitive statement can be made, variations in the Sun's brightness must be considered.

One of the main causes of short-term brightness variability is the presence of sunspots on the solar surface (Chapter 5). This is evident from the observed changes over the solar cycle (Chapter 6) where solar activity driven by the strong magnetic fields of sunspots varies by about 0.1% from solar maximum to minimum. The fact that the Sun is brighter when the coverage of the solar disk with sunspots is highest may seem counterintuitive given that sunspots are dark, and so more sunspots would suggest a less bright Sun. However, as discussed earlier in this chapter, the accumulated brightness of the faculae more than makes up for the reduction in emission from the sunspots, leading to an overall brightening of the Sun at solar maximum.

A recent study by climate and solar scientists from the United States, Switzerland, and Germany assessed changes in solar brightness over the last millennium and related these changes to their impact on the global average temperature of the Earth. Using direct observations from a range of high-sensitivity spacecraft in operation since 1978, this study found that there is no sign of a net increase in brightness over the thirty years spanned, and the cyclical changes were too weak to explain the measured rate of global warming over the same period.

Observations of sunspot activity have been recorded routinely since Harriot, Scheiner, Fabricius, and Galileo in the early 1600s, allowing for an accurate determination of the state and variability of the Sun over this timescale. Determining changes in solar brightness over longer timescales requires a different measure. Fortunately, a number of alternatives have been discovered that act as proxies for the sunspot number. Proxies for solar activity, such as the concentrations of various radioisotopes (e.g., carbon-14, beryllium-10 and chlorine-37) in the ice sheets of Greenland and the Antarctic, provide records of solar activity that can go back 1,000 years and more (see Chapter 6). This is a result of the fact that enhanced solar activity

Figure 11.8 Breakup of Antarctic Larsen B ice shelf. Courtesy of NASA/Goddard Space Flight Center Scientific Visualization Studio.

(higher sunspot numbers) enhances the solar wind that shields the Earth from the high-energy cosmic rays that produce the isotopes. Thus, changes in the isotope concentrations are directly related to the changes in sunspot activity. Using data from the historical records as input to terrestrial climate models, one can assess the effect of solar brightness variability on the global temperature. Studies show that even though the average magnitude of the solar maximum has increased over the last four hundred years, its impact on global warming over this time is far less than required to explain the increase entirely on the Sun. Estimates suggest that, at best, changes in solar brightness contribute about 25% of the increase in average global temperature.

Human civilization depends on a stable and predictable climate, something we have relied on for the whole span of our time on the planet. Today, global warming is a reality that threatens to change the Earth in irreversible ways. Understanding the sources of this warming, both natural and anthropogenic, is crucial to developing strategies to adapt to, slow down, and perhaps even reverse the trends. A key factor in this is the role played by the Sun. The radiation from the Sun is the sole source of energy for the heating of the Earth's surface and atmosphere, and solar variability has the potential to impact the Earth's climate on a variety of timescales. The most extensive studies to date show that the solar component to the measured increase in the global averaged temperature of the Earth is no more than 25%, and that the bulk of the warming is a result of the increase of greenhouse gases in the atmosphere, mostly carbon dioxide, methane, and nitrous oxide, from a range of current and past human activity. The ability of the Sun to grow vegetables is being severely tested by the needs of modern civilization.

RECOMMENDED READING

Fagan, Brian M. *The Little Ice Age: How Climate Made History, 1300–1850*. New York: Basic Books, 2000.

Lean, Judith, and David Rind. "Sun-climate Connections: Earth's Response to a Variable Sun." *Science*, 292(2001): 234–236.

Vogt, Gregory. *The Atmosphere: Planetary Heat Engine*. Minneapolis: Twenty-First Century Books, 2007.

WEB SITES

Climatology Basics: http://nsidc.org/arcticmet.

Geological history of Earth: http://www.fossilmuseum.net/GeologicalHistory.htm.

Greenhouse gases: http://www.eia.doe.gov/bookshelf/brochures/greenhouse/Chapter1.htm.

Intergovernmental Panel on Climate Change: http://www.ipcc.ch.

Science of Global Warming: http://www.ucsusa.org/global_warming/science/science-of-global-warming.html.

The Sun and Climate, Judith Lean and David Rind, 1996, Consequences, Volume 2, No. 1, http://www.gcrio.org/CONSEQUENCES/winter96/sunclimate.html.

12

The Future

As we head full thrust into the twenty-first century, we learn more and more about the universe every day. Yet despite four centuries of directly observing the Sun and solar phenomena, the small yellow dwarf star on our doorstep, the nearest and best studied astrophysical object to the Earth, is still very much a mystery. Observations of sunspots and their cycles, the discovery of a hot solar atmosphere, the ability to probe the interior of the Sun, and the elucidation of the Sun and Earth as an integrated system are all major milestones in our quest to understand this small corner of the universe. For every question answered, a hundred new ones appear.

The most recent decades have seen us conquer space, where specially designed telescopes allow us to explore the many facets of solar activity and variability. When combined with the dedicated observations from the ground, we have been able to build up a multidimensional picture of the Sun, covering wide ranges of timescales, length-scales, and wavelengths, with unprecedented detail. Currently, there are over a hundred ground-based observatories comprising more than 180 instruments worldwide, and more than twelve spacecraft in operation dedicated to observing the Sun. Much of our current knowledge of the Sun comes from these observatories, building upon past discoveries. However, despite the great advances made over the recent decades and the huge wealth of data gathered about every aspect of the Sun, there are still many unanswered questions.

OUTSTANDING ISSUES

Modern-day understanding of the Sun has developed to the point where the questions being asked are far more sophisticated than they were in the past. We have a rudimentary knowledge about most aspects of the Sun and the interactions between its various constituent parts, but some fundamental issues remain:

- *Generation of magnetic field and its consequences (Chapters 2, 3, 4, 5, and 6)*
  The science of helioseismology has opened up new doors to our understanding of magnetic fields in the Sun. While not complete, dynamo theory for the solar interior provides an extremely good description of how magnetic fields are generated. Questions remain, however, about the range of depths where generation of field occurs, whether or not a second more chaotic generation of magnetic field is required near the surface to explain the behavior of the magnetic carpet, what role the turbulent motions of the convective zone play in modifying the large-scale field that ultimately emerges through the surface to form sunspots, and how the solar dynamo mediates a complete reversal of the solar magnetic field. Moreover, the Sun is only one star, and it is not clear whether the theories and models developed for the Sun are relevant to other stars. The successful application of solar dynamo theories to Sun-like and non-solar stars will be the true test of this fundamental stellar process.

- *Understanding the chromosphere (Chapters 7 and 8)*
  Much focus over the last few decades has been centered on what heats the corona. Why is the corona so hot, and why is the heating structured in the way it is? However, whatever energy is available to heat the corona must have been transported through the underlying chromosphere. The shear mass of the chromosphere means that the amount of energy required to heat this region to the more modest 50,000 degrees is orders of magnitude larger than that required to heat the relatively thin corona to a few million degrees. The physics of the chromosphere is, to date, not well understood, and the complexity of the physical processes that occur, the wide range of temperatures and densities that exist, and the multiscale nature of its structure, make this a major challenge for the future. Without understanding the chromosphere, we cannot say that we understand the solar atmosphere.

- *Coronal heating (Chapter 7)*
  Ever since the discovery of highly ionized iron emission from the solar atmosphere, the mystery of the heating of the solar corona has been a perennial goal for solar physics research. It is clear that the energy is in the magnetic field, but how this energy is stored and how it is ultimately released is as yet unknown. Two major questions need to be answered: How is the magnetic energy dissipated released, and what determines the structure and distribution of the consequent heating?

- *Predicting solar flares and coronal mass ejections (Chapters 9 and 10)*
  Despite the major advances in recent years in understanding the physics and consequences of solar flares and CMEs both from the observational and modeling perspective, we are still no nearer to predicting when and where they will occur. This is a crucial component of the space weather enterprise, which, if successful, would allow accurate advance warning of potential radiation hazards for satellite operators, astronauts, and electrical grid technicians. This is particularly important in an era where the United States is advancing the human exploration of space, with plans being discussed for a human presence on the Moon and manned missions to Mars. Not only would the ability to predict these events benefit space weather forecasting, but it would also illuminate the key physics leading to the rapid and catastrophic energy release of magnetic energy in the solar atmosphere.

FUTURE STUDY OF THE SUN

A means by which to address the outstanding issues discussed previously is to observe the Sun in new and increasingly better ways. The advances made in the last fifteen years from an array of ground- and space-based observatories have been astounding. Ground-based solar helioseismology networks, such as BiSON (Birmingham Solar Oscillations Network) and GONG (Global Oscillation Network Group), in conjunction with the space-based SOHO helioseismology telescopes, SOI (Solar Oscillations Investigation), VIRGO (Variability of solar IRradiance and Gravity Oscillations), and GOLF (Global Oscillations at Low Frequencies) have revolutionized our understanding of the solar interior (Chapter 4). A vast array of instrumentation on board a fleet of spacecraft measuring an array of wavelengths, particles, and even plasma and magnetic fields in space have brought new perspectives and a greater understanding of the solar atmosphere and solar wind, their variability, and their interaction with the Earth and the solar system in general. These include, in chronological order of launch date:

- *Voyager 1 and 2* – Launched in 1977, the *Voyager* Interstellar Mission is an extension of the *Voyager* primary mission that was completed in 1989 with the close flyby of Neptune by the *Voyager 2* spacecraft. This extended mission continues to characterize the outer solar system environment and search for the heliopause boundary, the outer limits of the Sun's magnetic field, and outward flow of the solar wind.
- *Ulysses* – Launched in 1990, the *Ulysses* mission was the first spacecraft to explore interplanetary space at high solar latitudes, orbiting the Sun nearly perpendicular to the plane in which the planets orbit. The spacecraft and spacecraft operations team are provided by the European Space Agency (ESA); the launch of the spacecraft, radio tracking, and data management operations are provided by NASA.

- *Yohkoh* – Launched in 1991, *Yohkoh*, was an observatory for studying X-rays and gamma rays from the Sun, a project of the Institute for Space and Astronautical Sciences, Japan. The spacecraft was built in Japan, but the observing instruments had contributions from the United States and Great Britain. *Yohkoh* ceased operation in December 2001.

- *Wind* – Launched in 1994, *Wind* studies the solar wind and its impact on the near-Earth environment.

- SOHO – *Solar and Heliospheric Observatory* – Launched in 1995, SOHO, a joint venture of the ESA and NASA, is a solar observatory studying the structure, chemical composition, and dynamics of the solar interior; the structure (density, temperature, and velocity fields) and dynamics of the outer solar atmosphere; and the solar wind and its relation to the solar atmosphere.

- *Polar* – Launched in 1996, *Polar* is the second of two NASA spacecraft in the Global Geospace Science (GGS) initiative and part of the ISTP Project. GGS is designed to improve greatly the understanding of the flow of energy, mass, and momentum in the solar-terrestrial environment, with particular emphasis on geospace.

- ACE – *Advanced Composition Explorer* – Launched in 1997, ACE observes particles of solar, interplanetary, interstellar, and galactic origins, spanning the energy range from that of keV solar wind ions to galactic cosmic ray nuclei up to 600 MeV/nucleon.

- TRACE – *Transition Region and Coronal Explorer* – Launched in 1998, TRACE observes the effects of the emergence of magnetic flux from deep inside the Sun to the outer corona with high spatial and temporal resolution.

- IMAGE – Launched in 2000, *IMAGE* studied the global response of the magnetosphere to changes in the solar wind. Major changes occur to the configuration of the magnetosphere as a result of changes in and on the Sun, which in turn change the solar wind. IMAGE used neutral atom, ultraviolet, and radio imaging techniques to detect and gather data on these changes. IMAGE ceased operations in January 2006.

- RHESSI – *Ramaty High Energy Solar Spectrometric Imager* – Launched in 2002, RHESSI studies solar flares in X-rays and gamma rays. It explores the basic physics of particle acceleration and explosive energy release in these energetic events in the Sun's atmosphere. This is accomplished by imaging spectroscopy of X-rays and gamma rays with unprecedented spectral, spatial, and temporal resolution.

These spacecraft were augmented by a number of Earth magnetospheric and atmospheric space missions, as well as missions to Mars and the outer planets. See http://science.hq.nasa.gov/missions/sun.html for a complete list.

Recently, two major new missions were launched designed to observe the Sun as never before. *Hinode*, Japanese for sunrise, is a Japanese Space Agency mission with significant contributions from U.S. and European partners, was launched in August 2006. *Hinode* carries a suite of instruments including the Solar Optical Telescope (SOT), X-ray Telescope (XRT), and the Extreme Ultraviolet Imaging Spectrometer (EIS), which together are observing the solar surface and atmosphere in unprecedented fashion. The XRT is providing new data on the hottest parts of the solar corona, in particular during solar flares, and is yielding new insight into the heating of the corona and the acceleration of the solar wind. The EIS allows for detailed measurements of dynamic motions at a range of temperatures in the solar corona, generating exciting results on fast flows and energy release in a range of solar coronal phenomena. Particularly exciting is the SOT experiment on *Hinode*. The SOT is the largest aperture, most advanced solar telescope flown in space, and is capable of observing the solar surface and chromosphere at the remarkable resolution of 56 km per camera pixel. This is equivalent to spotting a penny from a distance of over 20 km (13 miles). The SOT is changing our view on how the magnetic field in sunspots emerge and evolve, how the chromosphere and corona interact, and how energy is transported through the chromosphere to power solar atmospheric transients. Figures 12.1 and 12.2 give examples of the extraordinary data produced by *Hinode*. Detailed information on the *Hinode* mission can be found at http://solarb.msfc.nasa.gov/.

The Solar Terrestrial Relations Observatory (STEREO) is a U.S. mission that was launched in October 2006. The STEREO mission is comprised of a pair of identical spacecraft with one placed in a drifting orbit moving ahead of the Earth and one in a drifting orbit lagging behind the Earth. The drifts result in the spacecraft separating from the Earth at about 22.5 degrees per year (45 degrees per year between the two spacecraft). The unique aspect of the STEREO mission is that it provides simultaneous and essentially identical views of the solar corona from two different perspectives, like a pair of eyes, enabling depth perception, and a measure of the fully 3D structure of the solar corona. Each of the STEREO spacecraft have a compliment of imaging telescopes and instruments to measure magnetic and electric fields, and the properties of energetic solar wind particles, with a major focus on understanding the 3D structure and evolution of coronal mass ejections. The Sun Earth Connections Heliospheric Investigation instrument suite is comprised of imagers that observe low solar corona at extreme ultraviolet wavelengths, a pair of coronagraphs (Chapters 7 and 9) that are sensitive to the faint corona, and CME enhancements out to a distance of about 15 solar radii, and a heliospheric imager that makes observations of the corona from 12–318 solar radii, that is, it will detect CME disturbances out beyond the orbit of the Earth. In addition, STEREO has an array of instruments and instrument suites that are designed to measure the properties of the plasma, particles, and magnetic and electric fields in interplanetary space. The IMPACT (In situ Measurements of PArticles and CME Transients) investigation provides measurements of the solar wind electrons, interplanetary magnetic fields, and solar energetic particles. The PLASTIC (PLAsma and SupraThermal Ion Composition) instrument studies coronal-solar wind and solar wind-heliospheric processes. S/WAVES (STEREO/WAVES)

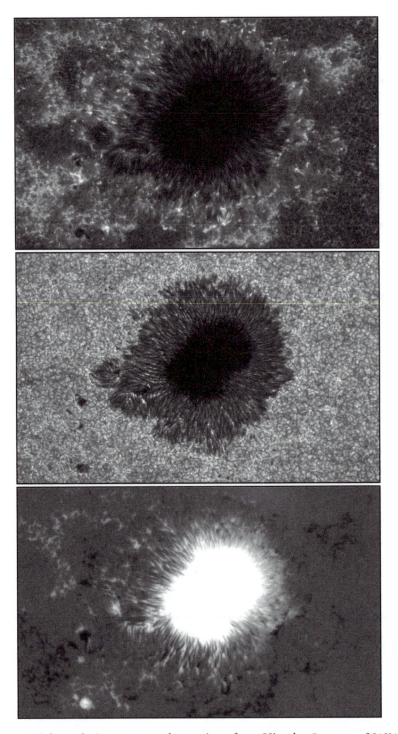

Figure 12.1 High-resolution sunspot observations from Hinode. Courtesy of JAXA and NASA.

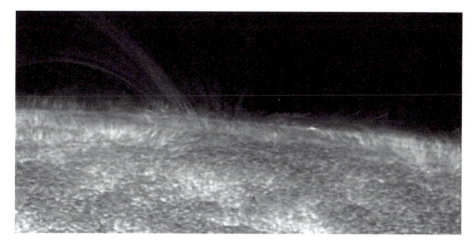

Figure 12.2 High resolution of the solar chromosphere from Hinode. Courtesy of JAXA and NASA.

is an interplanetary radio burst tracker that traces the generation and evolution of traveling radio disturbances from the Sun to Earth's orbit. With the STEREO mission we are observing the Sun from a unique perspective with startling new results. In particular, more information on STEREO can be found at http://stereo.jhuapl.edu/.

These are exciting times for solar physics research. The current array of solar or solar-related observatories both on the ground and in space is providing a wealth of extraordinary data of a caliber not seen before. We are observing the Sun from new vantage points, probing the interior with increasing accuracy and precision, observing in wavelengths spanning the whole electromagnetic spectrum, detecting directly the solar-produced particles and plasma flows in interplanetary space, all with improved resolution in space and time. And it is only going to get better.

In October 2008, NASA launched the *Interstellar Boundary Explorer* (IBEX) with the scientific goal of revealing the global properties of the boundaries that separate our heliosphere from the local interstellar medium. IBEX will image the distant boundaries and determine the strength of the termination shock at the edge of the heliosphere (Chapter 8). It will also discover what happens when the solar wind interacts with interstellar space by observing how the solar wind flows in the region between the shock and the very edge of the solar system (the heliopause), and how the interstellar gas interacts with the heliopause.

In August 2009, NASA will launch the *Solar Dynamics Observatory* (SDO), the first flagstone mission of the Living with a Star program. SDO will consist of four telescopes, spanning eight different wavelengths, all observing the solar atmosphere simultaneously with a high spatial resolution (each telescope will be able to resolve structures on the scale of 700 km) and with a high time resolution (Chapter 8), an optical instrument to measure the surface magnetic field, and a solar irradiance monitor (Chapter 11) to determine the Sun's radiation output at different wavelengths. SDO will help us to understand the how and why of the Sun's magnetic field changes, how it is generated and structured, and how the stored magnetic energy is released into the heliosphere. Measurements of the wavelength dependence of the solar irradiance (Chapter 11) will also help us develop the ability to predict the solar variations that influence life on Earth and humanity's technological systems.

Plans are also in place for the construction of the Advanced Technology Solar Telescope (ATST) funded by the National Science Foundation and due to be commissioned in 2012. The ATST will be the largest solar telescope in the world (4 m in diameter) and will be able to provide the sharpest views ever taken of the solar surface. The ATST will be able to see structures 70 km across, which is small enough to see the fundamental magnetic elements that make up the magnetic field pervading the solar surface.

Ideas for near-term solar facilities and missions also include COSMO (*Coronal Solar Magnetism Observatory*) to measure magnetic fields in the solar corona using infrared observations, and FASR (*Frequency Agile Solar Radiotelescope*) an ultrawideband radio array designed to make improved measurements of the nature and evolution of coronal magnetic fields and of the detailed physics of solar flares (Chapter 9). Ideas for near-term space missions include the European Space Agency's *Solar Orbiter* and NASA Sentinels missions. The *Solar Orbiter* will fly in an orbit that is a little more than half the distance of Mercury from the Sun (0.22 AU) and inclined to an angle of about 30°, which will allow measurements of the atmosphere above the solar poles. NASA's Sentinels mission is designed to provide the observations necessary for an understanding of the physics of the Sun and inner heliosphere. This is achieved with a combination of the Inner Sentinels, consisting of four identical spacecraft located close to the Sun, within 0.25–0.75 AU, primarily for in situ particles and fields measurements; a Near-Earth Sentinel with spectroscopic and wide field (>0.3 AU) coronagraphs, and a Farside Sentinel on the opposite side of the Sun from the Earth, with the ability to measure the magnetic field to provide, in conjunction with observations from the Earth, a view of almost the entire distribution of magnetic field around the Sun.

The legacy of 400 years of solar telescopic observation is continuing at a rapid pace with observational capability far beyond the dreams of Galileo. Future advances in our knowledge are assured by the wealth of ideas, the vision of solar scientists, and the active support of national science agencies worldwide.

EXTRASOLAR PLANETS

In 1995, a remarkable announcement was made by two Swiss astronomers, Michael Mayor and Didier Queloz. They claimed to have discovered a planet orbiting a Sun-like star. Planets had been discovered earlier, around exotic stars, but this was the first confirmed detection of a planet around an ordinary main sequence star. The star, known as 51 Pegasi, lies 50.1 light-years away in the constellation Pegasus. The planet was discovered via the detection of gravitational tugs exerted by the planet on the star, with the magnitude of the tugs suggesting a planet with a mass close to that of Jupiter, but lying as close as 8 million kilometers to the star (Jupiter is more than 778 million kilometers from the Sun). This now-famous discovery led to the development of improved spectroscopic and other techniques, with extrasolar planets being discovered thick and fast in the intervening thirteen years.

At the time of writing this volume, 342 individual planets orbiting other stars have been discovered by a variety of astronomy planet-finding teams around the world, and by a variety of advance detection methods. These 342 planets are part of 289 distinct planetary systems of which 37 have two or more planets. Most extrasolar planets discovered to date are of the size of Saturn or Jupiter, lying relatively close to their parent star—these are the easiest planets to find. There are some interesting examples with masses closer to that of the Earth or lying within a star's habitable zone. The number of known extrasolar planets changes daily. A full up-to-date catalog of all extrasolar planets can be found at http://planetquest.jpl.nasa.gov.

To date, our technology is not sufficiently advanced to routinely detect planets around other stars directly, although there is one confirmed direct planetary detection around a red dwarf star. Instead, indirect effects have been utilized to discover all of the 342 extrasolar planets known to exist. There are three main methods by which planets are discovered: gravitational wobble of the parent star, transit of star by the planet, and gravitational microlensing. Almost all extrasolar planets have been discovered by detecting small regular movements of the parent star, either by measuring the side-to-side motion of the star caused by the gravitational pull of a large planet orbiting close to the star (astrometric technique), or by measuring small regular and repeatable Doppler shifts in the spectral lines of the star's spectrum, signifying that the star moves in a small orbit of its own caused by the presence of a nearby mass, assumed to be a planet (Doppler technique). Almost 90% of all extrasolar planets discovered have been found by the Doppler technique. To illustrate how refined these measurements have to be, it is worth pointing out that the gravitational impact of Jupiter (orbital distance about ~5 AU) on the Sun as seen from a distance of ten light years (the typical distance of some nearby solar-like stars) would result in a side-to-side motion of the star of about 3 milli-arcseconds or about the size of an ant seen from a distance of 170 km. The further the star is from us, the smaller the shift observed. Using the Doppler technique, the same results would mean that a successful detection would require the ability to measure shifts in the wavelength of 1 part in 300 million.

The planet Earth causes gravitational motions of the Sun some thirteen times smaller than those of Jupiter. Therefore, these techniques

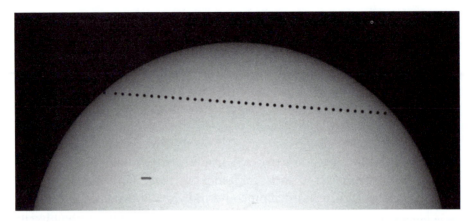

Figure 12.3 Transit of Mercury as seen by TRACE. Courtesy of NASA.

preferentially detect the *largest* planets, with the *closest* orbits to their parent stars, around the *nearest* stars. As the techniques become more refined and observations improve, smaller planets, planets further from their star, and planets around more distant stars will be detected. This is already evident over the course of the fourteen years since the first planet was discovered by Mayor and Queloz.

Another indirect way to infer the presence of a planet around a distant star is to observe the planet passing in front of the star. When such a planetary transit occurs, the planet blocks some of the light from its star, resulting in a periodic dimming. This periodic signature is used to detect the planet and to determine its size and orbit. We see this in our own solar system with both Mercury and Venus periodically transiting the Sun (Figure 12.3). To date, thirty-six extrasolar planets have been detected via the transit technique.

Missions to Detect Extrasolar Planets

The extrasolar planets that have been discovered so far are mostly gas giants, like Jupiter and Saturn, and as such are unlikely to support life as we know it. But some of these planetary systems might be expected to also contain smaller, terrestrial planets like Mars and Earth. NASA is embarking on a long-term enterprise to discover Earthlike planets around other stars, and to determine if they have the capability to support life.

Kepler, a NASA Discovery mission, is a spaceborne telescope designed to survey distant stars to determine the prevalence of Earthlike planets. Kepler, planned for launch in Spring 2009, will detect planets indirectly, using the "transit" method discussed in the main text. Kepler will hunt for planets using a specialized one-meter diameter telescope called a photometer to measure the small changes in brightness caused by the transits.

SIM PlanetQuest will follow on the heels of Kepler, to measure the distances and positions of stars with unprecedented accuracy, enabling the discovery of planets in the habitable zones around nearby stars. Currently, SIM PlanetQuest is slated for launch in the 2016–2020 timeframe. Finally, the Terrestrial Planet Finder (TPF), a suite of two complementary observatories, will be flown to study all aspects of planets outside our solar system. This includes their formation and development in disks of dust and gas around newly forming stars, to the presence and features of those planets orbiting the nearest stars, including an assessment of their suitability to support life. The basic idea of the TPF mission is to block the light from a parent star to see its much smaller, dimmer planets. This is an ambitious mission with major technological challenges that has been likened to finding a firefly near the beam of a brilliant searchlight from a large distance away. While scientists and engineers are optimistic that these challenges can be met, the mission is currently postponed awaiting future funding opportunities.

Our understanding of the Sun, the physics that drive its behavior, and its connection to the Earth both now and throughout history will continue to grow as our instruments, theories, and ideas improve. Through this understanding, we will learn more about other astrophysical bodies in the universe and, in particular, other stars and their planetary systems. The excitement of detecting other planets in the galaxy and the possibility that we may yet discover an Earthlike planet around a Sunlike star, perhaps even with signs of life, for example, atmospheric oxygen or water, has profound implications, not only for humanity, but also to some extent for our relationship to the Sun. The effect of solar variability on climate stability, habitability, and the evolution of life on Earth can then be compared with similar systems in other parts of the universe. One day we may even discover that the Sun is not the only star that grows vegetables.

WEB SITES

Advanced Technology Solar Telescope (ATST): http://atst.nso.edu.
Coronal Solar Magnetism Observatory (COSMO): http://www.cosmo.ucar.edu.
European Space Agency Solar Orbiter: http://www.orbiter.rl.ac.uk.
Extrasolar Planets: http://planetquest.jpl.nasa.gov.
Frequency Agile Solar Radiotelescope (FASR): http://www.fasr.org.
Interstellar Boundary Explorer (IBEX): http://ibex.swri.edu.
NASA Living with a Star (LWS): http://www.lws.nasa.gov.
Solar Dynamics Observatory (SDO): http://www.lws.nasa.gov/missions/sdo/sdo.htm.

Appendix: Interesting Facts about the Sun

- Light from the Sun takes 8 minutes and 19 seconds to reach the Earth.
- One cubic meter of gas from the solar corona at one solar radius above the surface contains less particles than a similar volume of most laboratory vacuums on Earth.
- The temperature of the solar corona is 300,000 times hotter than that of the photosphere.
- Every second the Sun emits enough energy to provide the entire United States with power for 4 million years.
- Every second the Sun loses approximately 4 billion tons of mass—the mass of Mount Everest every 4 minutes.
- The Sun is approximately 28,000 light-years from the center of the Milky Way galaxy.
- The Sun takes approximately 220 million years to orbit the center of the Milky Way galaxy.
- The Sun is the only star known to support life.
- The Sun is thought to be ~30% brighter today than it was when the Earth formed.
- The Sun is approximately 4.5 billion years old and is halfway through its life as a main sequence star.
- To shine, the Sun converts 600 million tons of hydrogen into helium in its core every second.
- A photon produced in the Sun's core takes approximately 300,000 years to reach your eye.

Glossary

Absorption lines. Dark lines superimposed on a bright continuum. The dark lines result from the absorption of light by cool material lying between the observer and the hotter source of the continuum light. Each absorption line lies at a specific wavelength defined by the properties of the elements making up the absorbing material.

Active latitudes. Range of latitudes on the Sun, typically 45° S to 45° N, in which sunspots develop and form; also known as the activity belt.

Active longitudes. Specific longitudes on the Sun associated with the recurrence of flare-productive active regions.

Active nests. Localized regions on the Sun in which a succession of sunspots appears over time.

Active region prominences. A solar prominence occurring in an active region. Typically, an active region prominence is low-lying, ~5000 km, and displays significant dynamic motions within it.

Active Sun. Regions of the Sun that display enhanced emission and activity. These regions are generally regions of strong magnetic field sometimes containing sunspots.

Albedo. Diffuse reflectivity of an object. Typically, the term is used to refer to the fraction of sunlight reflected from a moon or planet in the solar system.

Alpha-effect. Important process in the generation of a magnetic field on the Sun. Solar rotation results in the twisting of magnetic field lines as they make their way to the surface through the convection zone. The twisted field lines look like the Greek letter α.

Antinode. The point on a wave where the maximum amplitude is achieved.

Arcade. An array of hot solar loops lying in succession along a magnetic neutral line. An arcade is exemplified by the magnetic structure in the aftermath of a solar flare.

Astrolabe. An ancient astronomical calculator used to determine time and position of the Sun and the stars.

Astrometric technique. One of the methods by which planets are discovered around other stars. The astrometric technique uses precise measurements of the parent star's position to detect the slight wobbles due to the gravitational influence of an orbiting planet.

Astronomical unit (AU). The mean earth-sun distance, equal to 1.496×10^8 km.

Asteroseismology. The study of oscillations on stars other than the Sun to determine the properties of the star's internal structure. Using similar techniques to helioseismology, the frequency spectra of the pulsating star is analyzed to determine the effect of the oscillations.

Aurora. Faint lights, seen mostly in the nighttime sky at high latitudes, associated with geomagnetic activity. Auroras occur typically occur at heights between 100 to 250 km above the ground.

Auroral bands. Flickering ribbon or sheet-like structures appearing in the aurora.

Auroral electrojet. An electrical current that flows in the ionosphere in the auroral zone, associated with geomagnetic activity.

Babcock model. A model proposed by Horace Babcock in 1961 to explain the observed patterns displayed by sunspots and the Sun's magnetic field.

Bartels diagram. Visual representation of geomagnetic activity introduced by Julius Bartels in 1949. The Bartels diagram shows the quantitative values of the geomagnetic index, *Kp*. Its main feature is its division into blocks denoting individual solar rotations, allowing for easy identification of recurring geomagnetic storms.

Blazars. A very compact energetic object associated with supermassive black holes at the centers of galaxies.

Bremsstrahlung radiation. Electromagnetic radiation produced when a charged particle is deflected by another charged particle. Typically this involves high-energy electrons being deflected by slow-moving ions to produce high-energy photons.

Carbon-14. Radioactive isotope of carbon, sometimes referred to as radiocarbon.

Chromosphere. Region of the Sun immediately above the surface (or photosphere). The chromosphere is defined by temperatures ranging from 10,000–100,000 K.

Conduction. Process describing the transfer of thermal energy, or heat, through matter, from a region of higher temperature to a region of lower temperature.

Conductive flux. The amount of thermal energy transmitted per unit area per second.

Conservation of energy. Fundamental concept of physics stating that energy can be neither created nor destroyed. Energy can only change between different types, for example, magnetic to thermal, or potential to kinetic.

Continental shelf. Extended area of shallow seas surrounding each continent and associated coastal plain.

Convection. Process by which energy and mass is transferred through a fluid. The warmer portions of the fluid are less dense and, therefore, rise, while, the cooler portions are more dense and, therefore, fall.

Convection zone. Region of the Sun taking up the outer 30% of the solar interior. Energy transport in this region is dominated by convection.

Copernican revolution. Refers to the transition from a viewpoint in which the Earth was the center of the solar system, called geocentric, to one in which the Sun was at the center, called heliocentric; named after Nicolas Copernicus, who promoted the idea in his 1543 book *De revlutionibus.*

Corona. Outer region of solar atmosphere, named after its crown-like appearance during solar eclipses.

Cosmic microwave background (CMB). Relic radiation from the Big Bang. Discovered in 1965, it provides strong evidence for the Big Bang theory of the universe. The CMB pervades the whole universe and has a present-day temperature of around 2.7 K, having cooled over the lifetime of the universe from its very hot origins.

Cosmic rays. Energetic particles traveling through space and interacting with the Earth's atmosphere. Thought to originate in neutron stars, black holes, and supernovae, cosmic rays are mostly composed of protons, with about 9% helium nuclei and 1% electrons. Energies of these particles can exceed 10^{20} eV.

Cross-staff. A measurement tool that was widely used by astronomers and navigators before the invention of the telescope. It consists of a main staff with a movable perpendicular crosspiece attached symmetrically to the staff. A cross-staff was used to measure the angle between the directions of two stars.

De-excitation gamma-ray line spectrum. Collection of distinct emission lines in the energy range 4–8 MeV observed in solar flares and resulting from the relaxation of ambient ions in the solar atmosphere to a lower energy. The ions are originally excited to higher energies by collisions with accelerated protons and helium nuclei.

DeVries Cycle. A 205-year cycle of solar variability suggested by the observed radiocarbon periodicities in tree ring studies.

Differential rotation. When different parts of a rotating object, like the Sun, rotate with different angular rates of change, the object is said to be differentially rotating.

Dimming. Sudden reduction in soft X-ray or EUV emission associated with eruptive flares and coronal mass ejections.

Disparition brusques. A solar filament or prominence that disappears suddenly within a few minutes or hours. The phenomenon was first studied by French solar astronomers, hence the name.

Doppler technique. The primary method by which planets are discovered around other stars. The Doppler technique uses measurements of the shifts in the wavelengths of the light from the parent star to determine radial velocity of the star as it wobbles under the gravitational influence of an orbiting planet.

Dynamo. Generally, a dynamo is a machine that converts mechanical energy into electrical energy, which, in turn, can result in the generation of a magnetic field in a conducting medium like the ionized plasma of the solar interior.

Ecliptic plane. Plane centered on the Sun that contains the orbit of the Earth. From the perspective of the Earth, the ecliptic is the path of the Sun across the sky throughout the year.

Electromagnetic radiation. Radiation propagating at the speed of light that results from the motion of electric charges. The most common form of electromagnetic radiation is visible light.

Electromagnetism. The study of the interaction between electric and magnetic fields.

Electron-Volt. The energy gained by an electron when accelerated through an electrical potential difference of 1 V. The electron-Volt is the primary unit of energy for particles in the solar atmosphere.

Electro-optic modulator. An optical device used to modulate a beam of light via the manipulation of electric fields.

Emission line. Bright line resulting from the emission of light by a hot gas. Each emission line lies at a specific wavelength defined by the properties of the elements emitting the light.

Epicycles. Fictitious circle about a point on a planetary orbit. Epicycles were introduced by Ptolemy in his Earth-centered solar system to account for the observed motion of the planets while restricting them to move in perfect circles with uniform velocity; also used by Copernicus in his Sun-centered system.

Equinox. Times during the year when the Sun is directly over the Earth's equator and day and night are of equal length.

Evershed effect. Strong horizontal outflow observed in the penumbra of sunspots, extending several thousands of kilometers from the edge of the sunspot umbra.

Excited state. State of an atom with an energy higher than its lowest allowable energy. The lowest allowable energy for a given atom is known as its ground state.

F-corona. That part of the corona seen during a solar eclipse that is due to sunlight scattered or reflected by dust in interplanetary space.

Faculae. Bright regions seen around sunspots and occurring a few hundred kilometers above the photosphere. Faculae are mostly seen near the solar limb, although they occur all across the Sun.

Filament. Cool dense chromospheric material supported in the corona by a magnetic field. Seen in absorption as dark elongated structures on the solar disk in Hα observations. At the solar limb, filaments are seen as bright structures in emission and are then known as prominences.

Five-minute oscillations. Pulsations in the solar photosphere with a five-minute period. The strongest oscillation exhibited by the Sun, and the first discovered, was in 1962.

Free radical. A highly reactive molecule with an unpaired electron. Free radicals are dangerous; they can do a lot of damage when they interact with cellular material and DNA.

Fundamental frequency. The lowest frequency of a vibrating system. For sound, the fundamental frequency determines the "pitch" of a tone.

Gamma-ray bursts. Bursts of radiation at gamma-ray wavelengths, emanating from all directions. Gamma-ray bursts are hundreds of times brighter than a typical supernova and about a million trillion times brighter than the Sun.

Gamma-ray continuum. Diffuse emission at gamma-ray energies devoid of any emission or absorption lines; dominated by emission from the galactic center.

Geo-effective. Having an impact on the Earth's magnetic environment.

Geomagnetic storms. A strong enhancement in the Earth's magnetic activity, with a strength measured by large values of the geomagnetic indices, most notably the Ap index.

Geomagnetic tail. Region of the Earth's magnetosphere on the anti-sunward side, stretched out to more than 200 Earth radii by the solar wind.

Geosynchronous orbit. A circular orbit about the Earth at a radius of 42,300 km above the equator and with an orbital period of one day. A satellite in such an orbit remains directly above the same point on the Earth.

Gleissberg Cycle. An 80-year modulation cycle exhibited by the amplitude of the eleven-year sunspot cycle.

Global modes. Oscillation signatures of the whole Sun. Each mode represents a specific amplitude of pulsation in each of the three radial, azimuthal, and longitudinal directions making up the sphere.

Golden Record. A vinyl phonograph record carried onboard the two *Voyager* spacecraft launched in 1977. Each record contains sounds and images portraying the diversity of life and culture on Earth.

Granulation. Cell-like structure visible on the solar photosphere, resulting from convective motions.

Gravitational constant. The universal constant in Sir Isaac Newton's theory of gravity, denoted by the capital letter G. It has a value of $G = 6.67 \times 10^{-11} \, m^3 \, kg^{-1} \, s^{-1}$.

Greenhouse gas. A gas that absorbs infrared radiation in the Earth's atmosphere. Common greenhouse gases are water vapor, carbon dioxide, and methane.

Ground level event (GLE). Dramatic increase of neutron counts on the surface of the Earth as a result of energetic particles from large solar flares making it through the magnetosphere and interacting with the Earth's atmosphere. Only a few GLEs occur in a solar cycle.

Habitable zone. A region of space surrounding a star where the temperature is favorable for life to exist.

Hale-Nicholson Polarity Law. Rule followed by solar active regions such that, in a given solar cycle, active regions in the northern hemisphere have a positive-negative (or negative-positive) polarity pairing, relative to the direction of the solar rotation, while this pairing is opposite in the southern hemisphere. The orientation of the pairing alternates between odd- and even-numbered solar cycles.

Hallstatt Cycle. Proposed 2300-year cycle of solar activity coinciding with marked periodicities in the carbon-14 record in tree rings and ice cores.

Heliopause. Boundary between the heliosphere and the interstellar medium.

Helioseismology. The science of understanding the pulsations of the Sun to yield information on the structure and dynamics of the solar interior.

Heliosphere. The region dominated by the solar wind and marking the extent of the Sun's influence.

Heliospheric current sheet. The structure that separates the regions of positive and negative magnetic polarity in interplanetary space. The structure the heliospheric current sheet forms is defined by the combination of solar rotation and the solar wind, and takes on the appearance of a "ballerina skirt."

Helmet streamer. Structure seen in the solar corona comprising a central core of closed magnetic loops overlaid by radially stretched open field, giving the appearance of a helmet.

HK Project. A project of the Mount Wilson Observatory near Pasadena, California, which has been observing the light emitted by calcium ions in the chromospheres of active magnetic regions in nearby stars. The H and K in the name refer to the specific lines in the spectrum of calcium ions.

Holocene. A period in geology beginning approximately 11,500 years ago and which continues until the present day. The Holocene is part of the Quaternary period.

Hydrogen burning. The nuclear fusion process whereby four protons (or hydrogen nuclei) interact to form one helium nucleus. Hydrogen burning occurs in the core of main sequence stars, and the energy released maintains the star against gravitational collapse.

Hydrostatic atmosphere. An atmosphere in which the force due to pressure gradients in the gas is exactly balanced by gravitational forces to maintain an equilibrium state.

Interference. The effect caused by the combination of two interacting waves. Depending upon the amplitudes and frequencies of the waves, they can either add to each other (constructive interference) or cancel each other (destructive interference). This principle applies to all waves: electromagnetic, sound, and water waves.

Interplanetary magnetic field (IMF). The magnetic field that fills the heliosphere. The field is anchored in the Sun, which rotates, and is carried out into interplanetary space by the solar wind. The result is a spiraling magnetic field separated into a northern and southern hemisphere by the heliospheric current sheet.

Interplanetary medium. The space permeated by the solar wind and spanning the whole solar system.

Interstellar medium. The diffuse distribution of gas and dust that fills the space between stars and galaxies.

Ion. An atom that has lost or gained one or more electrons giving it a positive or negative electrical charge.

Isotope. Atoms of the same element that have the same number of protons, but different numbers of neutrons.

Joy's Law. A rule generally obeyed by solar active regions, which states that the leading polarity, with respect to solar rotation, will lie closer to the equator than the following polarity.

K-corona. The part of the corona seen during a solar eclipse that is due to sunlight from the photosphere scattered by coronal electrons.

Kelvin. Unit of temperature on the absolute temperature scale proposed by physicist Lord Kelvin. Zero Kelvin is referred to as absolute zero, corresponding to −273.15 degrees Celsius.

Lagrangian point. One of five locations within the Sun-Earth system where a small object can be stationary against the gravitational pull of both the Earth and the Sun. Of particular importance is the L1 point, which lies at the gravitational balance point along the Sun-Earth line, approximately 1.5 million kilometers from the Earth.

Last scattering surface. The physical region, assumed to be the surface of a sphere, in the early history of the universe at the point when the universe became transparent to light. Before this time, photons scattered readily from free electrons since the temperature was high enough to ionize hydrogen. After the temperature cooled, the electrons were trapped in hydrogen atoms and so the light was free to stream through the universe. This is similar physics, which defines the sharp edge of the solar photosphere.

Light-year. Distance traveled by light in one year and equal to 9.46 trillion kilometers.

Lodestone. Naturally occurring magnet, also known as magnetite.

Longitudinal wave. A wave in which the propagation of the signal is in the same direction as the wave motion. A sound wave is a longitudinal wave.

M regions. Unobservable regions on the Sun assumed responsible for recurring geomagnetic storms. They were discovered in the late 1970s to be equatorial extensions of coronal holes.

Magnetic bipoles. Positive-negative pair of magnetic polarities observed on the solar surface.

Magnetic buoyancy. The mechanism by which a magnetic field rises through the convection zone to break through the photosphere; result of strong magnetic pressure in magnetic flux tubes.

Magnetic butterfly diagram. Latitude versus time relationship showing behavior of Sun's magnetic field over multiple solar cycles; name comes from the characteristic shape of butterfly wings that appears as a new field emerges closer to the equator as the cycle progresses.

Magnetic declination. Measurement at a given geographic location of the azimuthal component of the Earth's magnetic field vector relative to geomagnetic north.

Magnetic dipole. Magnetic bipole where the opposite polarities have the same strength of magnetic field.

Magnetic field lines (or, magnetic lines of force). Magnetic field lines describe the shape of magnetic fields in three dimensions, defined by the direction a compass needle points. Field lines converge where the magnetic force is strong, and spread out where it is weak.

Magnetic flux density. The amount of magnetic flux per unit area passing through a given region lying perpendicular to the direction of the flux; specified by the number of magnetic field lines contained within a given area.

Magnetic flux tube. Concentrations of magnetic field into elongated tube-like structures.

Magnetic helicity. Conserved property of a magnetic field in closed magnetic volumes; defines the amount of twist in the magnetic field lines.

Magnetic loops. Structure in the solar corona in which certain magnetic field lines are filled with hot emitting plasma. Coronal loops are rooted at each end in the photosphere.

Magnetic neutral line (or, polarity inversion line). The location on the solar surface where the line-of-sight magnetic field changes polarity from outward to inward, or vice versa.

Magnetic polarity. Magnetic fields can have a positive or negative polarity signifying whether the field is pointing outwards or inwards on a surface.

Magnetic pressure. Pressure exerted by magnetic fields, described by the tendency of neighboring magnetic field lines to repulse each other.

Magnetized plasma. Plasma in which the ambient magnetic field is strong enough to significantly alter the paths of the particles in the plasma.

Magnetograph. An instrument for measuring solar magnetic fields providing maps of the magnetic field strength and/or direction.

Magnetopause. The boundary of the Earth's magnetosphere, separating the plasma of the magnetosphere from that in the solar wind.

Magnetosphere. The region of space around a planet containing the planet's magnetic field, and confined by the solar wind.

Main sequence. Name given to a strong band of stars on a plot of brightness versus temperature. Most stars we can observe lie on the main sequence and spend most of their "lives" there. Main sequence stars are in their hydrogen-burning phase of energy production.

Mantle. Layer of an Earth-like planet composed of medium-density rock lying below the lighter density crust.

Mantle convection. Heat transfer process in the mantle where hot rock from below slowly replaces cool rock above.

Meridional flow. General flow pattern along lines of longitude from equator to poles in the Earth's atmosphere or the solar interior.

Milankovitch Cycle. Collective name for cyclic variations observed in the Earth's orbital motion thought to have an impact on the periodic changes in the Earth's climate. These variations consist of changes in the eccentricity of the Earth's orbit, in the tilt of the Earth's rotation axis, and the precession of the orbit; named after Serbian astronomer Milutin Milankovitch.

Nanoflares. Small energy releases in the solar corona thought to be responsible collectively for heating coronal loops.

Neutrino. Stable elementary particle of small mass and no charge that interacts weakly with normal matter making the particles difficult to detect. There are three types of neutrino, the electron-neutrino, muon-neutrino and tau-neutrino, named after the particles they pair with in their production.

Nodal lines. Lines passing through all the nodes marking the locations of zero amplitude when two waves interfere.

Node. A point of zero displacement on a standing wave.

Omega-effect. Mechanism by which differential rotation can translate a north-south directed, or poloidal, magnetic field to east-west directed, or toroidal, magnetic field, thereby facilitating the dynamo process.

p-p chain. The proton-proton, or p-p, chain describes the sequence of reactions in nuclear fusion that leads to the production of a single helium nucleus from four protons.

Parker spiral. The shape of the Sun's extended magnetic field derived from a combination of radial stretching by the solar wind and winding by solar rotation; named after astrophysicist Eugene Parker.

Penumbra. Lighter region of a sunspot surrounding the umbra and extending 7,000–10,000 km.

Photon. Elementary particle of zero mass representing the smallest bundle of energy of light at a particular frequency.

Photomultiplier. Sensitive light detectors in the ultraviolet, visible, and near infrared wavelengths. Light signals can be amplified by a factor of greater than 100 million. Incoming photons generate a cascade of secondary electrons that is detected as a current.

Photosphere. Visible surface of the Sun.

Planck constant. Denoted h, this defines the elementary quantum of action relating frequency of light to its energy. Planck's constant has a value of $h = 6.626 \times 10^{-34}$ Joule-seconds.

Planetary nebula. The ejected outer envelope of a low-mass star that has completed its hydrogen burning phase.

Planetary transit. Passage of a planet across the face of its parent star, resulting in a blocking of a fraction of the light from the star depending upon the size of the planet, the distance it is from the star, and the inclination of its orbit relative to the direction of the observer.

Plasma. An ionized gas comprised of ions and electrons.

Plate tectonics. Theory of how sections, or "plates," of the Earth's crust and upper mantle form, move, and interact.

Pleistocene. The first phase of the Quaternary Period, immediately preceding the Holocene era. The Pleistocene is thought to range from about 1.8 million years ago until about 11,500 years ago, and covers much of the period of ice ages on the Earth.

Polar crown. Region around the poles of the Sun at about latitude 70° containing east-west oriented filaments and prominences. Often the initiation sites of coronal mass ejections.

Poloidal field. Component of Sun's magnetic field, which runs north-south from pole to pole.

Pores. Small umbral sunspots with no discernible penumbra; often associated with larger sunspots.

Positron. Antiparticle of the electron, with the same mass but opposite charge.

Postflare loops. Distribution of hot bright loop-like structures seen in the corona following a solar flare. Often forms a bright arcade in soft X-rays or EUV.

Prominence. Cool dense chromospheric material supported in the corona by a magnetic field. Seen in emission as bright structures at the solar limb in Hα observations. On the solar disk, prominences are seen as dark structures in absorption, and are then known as filaments.

Proton storm. The rapid enhancement of the measured proton flux above an energy of 10 MeV at the Earth. Proton storms can produce an array of geomagnetic phenomena, including an increased ionization of the ionosphere.

Quantum mechanics. A fundamental branch of physics describing the interactions between elementary particles and photons. Most quantum mechanical physics is only observable on the subatomic scale.

Quantum number. A number used to categorize conserved quantities in a dynamical system that is governed by quantum mechanics, such as energy of an atom.

Quaternary Period. The geologic period extending from roughly 1.8 million years ago to the present. Contains two distinct epochs: the Pleistocene and Holocene eras. It covers the time span of glaciations and their associated alluvial deposits.

Quiescent prominences. A prominence that shows little large-scale motion, develops very slowly, and has a lifetime of several solar rotations. Quiescent prominences form within the remnants of decayed active regions, in quiet areas of the Sun between active regions, or at high solar latitudes where active regions seldom or never form.

Quiet solar network. Localized regions of magnetic enhancement outside of sunspots and active regions.

Quiet Sun. Regions of the solar surface with no sunspots or prominences.

Radiation belts. Regions of space surrounding the Earth containing enhanced densities of energetic electrons or protons near the magnetic equator; often called the Van Allen belts after their discoverer.

Radiation sickness. Term describing the effects of damage to internal organs as a result of exposure to high doses of radiation.

Radiation zone. Interior region of the Sun surrounding the core and bordered above by the convection zone. Energy, generated in the core, is transferred through the radiation zone by the absorption and reemission of the energy by atoms.

Radiative energy. Amount of energy emitted as radiation.

Radiative flux. Amount of radiative energy flowing through a unit area every second.

Radioactive isotope. An isotope of a given element that is unstable and releases energy by spontaneously emitting radiation.

Radiocarbon. See carbon-14.

Red giant. The evolutionary phase of solar mass stars during which the primary energy source is hydrogen burning in a thin shell around a predominantly inert helium core.

Roentgen Equivalent Man (rem). Unit of radiation dose quantifying the biological effect of an absorbed dose of radiation.

Sextant. An astronomical and navigational instrument used to measure the altitude of a celestial object, like the Sun, above the horizon. A common use of the sextant is to sight the Sun at noon to find one's latitude.

Shear layer. A layer in a fluid across which there is a sharp change in the speed or direction of the flow.

Shock wave. Structure formed in a fluid flow when a disturbance propagates faster than the typical signal speed of the fluid. Shock waves are characterized by a sudden change in the properties of the fluid, such as pressure, density, or magnetic field strength.

Sieverts. Unit of radiation dose equal to the absorption of 1 Joule of energy per kilogram of the absorber.

Sigmoids. S-shaped structures in the solar corona observed in soft X-ray and EUV wavelengths. Sigmoids indicate magnetic fields with excess energy that may be released in a CME.

Single event upset (SEU). A change of state in sensitive electronic equipment caused by the impact of a low-energy ion or electron.

Solar day. The length of time from noon to noon as defined by the passage of the Sun across the sky. The solar day varies from the mean solar day of twenty-four hours by as much as seventeen minutes.

Solar dynamo. The process that governs the interaction of magnetic fields and plasma in the solar interior.

Solar energetic particles (SEP). Energetic ions and electrons accelerated in solar flares and coronal mass ejections. Some of the highest energy particles can reach 80% of the speed of light.

Solar limb. The apparent edge of the solar disk.

Solar maximum. The time at which the daily average sunspot count reaches its peak during a given solar cycle.

Solar minimum. The time at which the daily average sunspot count reaches is lowest value during a given solar cycle.

Solar nebula. The cloud of gas and dust that collapsed under its own gravity to form the solar system.

Solar proton events (SPE). Associated with solar flares and coronal mass ejections, a solar proton event signifies the acceleration of protons to very high energies that penetrate the magnetosphere to create enhanced geomagnetic effects.

Solar spectrograph. An instrument that spreads solar light or other electromagnetic radiation into its component wavelengths.

Solar wind flux. Amount of solar wind flowing through a unit area in one second.

Solstice. Times during the year when the Sun is at its maximum position south or north of the equator. The winter solstice occurs around December 21–22, while the summer solstice occurs around June 20–21 each year.

Space Age. Period in Earth's history defining the time over which humankind has been able to send personnel and equipment into space. The period started in 1957 with the launch of the Soviet satellite *Sputnik*.

Spectroscopy. The study of a source of electromagnetic radiation by dispersing the incoming radiation into a spectrum of different wavelengths.

Spicules. Dynamic jets of plasma seen in the solar chromosphere, extending as high as 11,000 km.

Spörer's Law. The rule generally followed by emerging sunspots whereby early in the solar cycle they appear at high latitudes ($\sim30°$ north or south), but then appear at increasingly lower latitudes as the cycle progresses. This rule is exemplified by the butterfly diagram.

Stefan-Boltzmann Law. Physical law that relates the power of a radiating object, like a star, to its temperature. The Stefan-Boltzmann Law states that the emission per unit area per second, j, of the object is directly proportional to the fourth power of the temperature, T:

$$j = \sigma T^4$$

where $\sigma = 5.67 \times 10^{-8}$ J m^{-2} s^{-1} K^{-4} is the Stefan-Boltzmann constant.

Stellar evolution. The set of processes followed by a star from birth to death; these process are driven by how quickly a star processes its hydrogen, helium, and heavier elements.

Stratosphere. The part of the Earth's atmosphere immediately above the troposphere, extending to a height of about 50 km.

Strong force. A fundamental force of nature that governs how protons and neutrons are held together in an atomic nucleus.

Subduction zone. A geological zone where cold, dense ocean crust is pushed beneath less dense continental crust. This typically occurs at the boundary between ocean and continental tectonic plates.

Sudden ionospheric disturbance (SID). An extraordinarily high ionization of the Earth's lower ionosphere caused by a solar flare.

Supergiant star. The evolved phase of a star that has over five times the mass of the Sun. Supergiants are more than 10,000 times brighter than the Sun and one billion times the volume.

Supergranulation. Large-scale pattern of convection seen on the solar surface.

Tachocline. An area of rapid change between the convection zone and radiative zone, occurring about one-third of a solar radius below the surface.

Temperature minimum region. Region in the solar atmosphere where the temperature reaches its minimum value of around 4,400 K. This occurs at a height of about 500 km above the surface.

Termination shock. The point where the solar wind becomes subsonic as it begins to feel the effects of the interstellar medium.

Thermal energy. Energy content of a volume of plasma related to its temperature.

Thermal pressure. The pressure of a volume of plasma arising from the motions of particles that result from the plasma's temperature.

Toroidal field. Component of Sun's magnetic field that runs east-west around the Sun, parallel to the equator.

Trade winds. Wind pattern found around the Earth's equatorial regions. The trade winds are the prevailing winds in the tropics blowing from the northeast in the northern hemisphere and the southeast in the southern hemisphere.

Transition region. Thin layer separating the solar chromosphere from the solar corona. The temperature varies from about 100,000 K to 1 million K over 100–500 km.

Tropical year. The length of time that the Sun takes to return to the same position in the sky as seen from Earth, that is, from winter solstice to winter solstice; also known as the solar year.

Troposphere. Layer of the Earth's atmosphere immediately above the planet's surface and rising to about 10 km. The troposphere is heated from below by infrared radiation from the sun-heated Earth.

Umbra. Central, dark portion of a sunspot.

White dwarf. A small, dense star that has exhausted most of its internal nuclear fusion resources. It is the final stage of evolution of a solar mass star.

White light. Solar radiation integrated over the visible portion of the spectrum (from 400 nm to 700 nm).

Wien's Law. Physical law that states the peak wavelength of emission from a radiating body, like a star, is inversely proportional to the temperature of the star.

Wolf number. A historic term for the sunspot number, a number that represents the number of sunspots on the solar surface; named after Rudolf Wolf, who originated the idea of counting sunspots as a measure of solar activity in 1949.

X-ray bright points. Small, compact, and short-lived enhancements at soft X-ray wavelengths, often seen in coronal holes.

Zodiacal light. Band of diffuse light seen near the ecliptic immediately after sunset or before sunrise due to the reflection of sunlight by interplanetary dust.

Annotated Bibliography

BOOKS AND REPORTS

Benestad, Rasmus E. *Solar Activity and Earth's Climate.* Berlin, Heidelberg, and New York: Springer-Praxis, 2006. (ISBN-10: 354030620X; ISBN-13: 978-3540306207)

> Good introduction to the role the Sun plays in climatic changes on the Earth. It details the observations and physical processes relating variations in the Sun's output over long and short timescales to their response at the Earth.

Bleeker, J.A., J. Geiss, and M. Huber, eds. *The Century of Space Science.* Berlin, Heidelberg, and New York: Springer-Verlag, 2002. (ISBN-10: 0792371968; ISBN-13: 978-0792371960)

> A collection of historical review articles documenting the major discoveries in space physics, solar physics, and astrophysics by some of the top scientists in their respective fields. Several chapters deal with the Sun, providing unique insight into the history of space-based science and the advancements made in our understanding of the Sun.

Carlowicz, Michael, and Ramon Lopez. *Storms from the Sun: The Emerging Science of Space Weather.* Washington, DC: National Academies Press, 2002. (ISBN-10: 0309076420; ISBN-13: 978-0309076425)

> This book provides an excellent description of the science of space weather and the issues associated with understanding the Sun's effect on the Earth. Written for a public audience, this volume details the principal effects of space weather and tells in colorful detail the stories associated with real solar storms.

"Climate Change 2007"—the IPCC Fourth Assessment Report (AR4), Synthesis Report: Summary for Policy Makers. Downloadable from http://www.ipcc.ch.

> A must read for anyone concerned about global warming. This summary synthesizes the reports of the various working groups of the Inter-governmental Panel on Climate Change (IPCC) showing the change in greenhouse gas composition in the Earth's atmosphere, and its association with the Earth's average temperature, together with a discussion of the sources of this change, the impact of this change, and how to mitigate the effects.

Golub, Leon, and Jay M. Pasachoff. *The Solar Corona.* Cambridge: Cambridge University Press, 1997. (ISBN-10: 0521485355; ISBN-13: 978-0521485357)

A comprehensive review of the hottest part of the solar atmosphere, the solar corona. It brings together ground-based and space-based observations to describe the physical processes involved in the heating and dynamics of the corona, and how our understanding of this enigmatic region of the solar atmosphere has developed.

Lang, Kenneth R. *The Sun from Space.* 2nd ed. Berlin, Heidelberg, and New York: Springer-Verlag, 2008. (ISBN-10: 3540769528; ISBN-13: 978-3540769521)

An extremely well-written book focusing on what we have learned about the Sun from observatories in space. Each chapter highlights a region of the Sun and discusses some of the discoveries made about each region, with great use of color illustrations showing real data.

Sobel, Dava. *Longitude: The True Story of a Lone Genius Who Solved the Greatest Scientific Problem of His Time.* New York: Walker and Company, 1995. (ISBN-10: 0007790163; ISBN-13: 978-0802715296)

Great read detailing the development of the first accurate chronograph enabling sailors to accurately calculate their longitude while at sea. Contains a detailed description of how the Sun can be used in navigation.

WEB SITES

These Web sites are indispensible:

http://planetquest.jpl.nasa.gov: The exoplanet exploration Web site of the NASA Jet Propulsion Lab in Pasadena, California. An up-to-the-minute survey of all planets discovered around distant stars.

http://www.spaceweather.com: The go-to place for current space weather conditions and tutorials on space weather physics.

http://science.nasa.gov: Science @ NASA Web site, highlighting the latest discoveries from NASA.

http://sprg.ssl.berkeley.edu/shine/suntoday.html: The Sun Today. Nice Web site linking to many of the ground-based and space-based solar observatories showing current views of the Sun at many different wavelengths.

Index

Absorption lines, 15–16, 23, 36, 96

Active regions, 106–7, 112, 150, 160–61; active longitudes and active nests, 91; active Sun, 63

Activity belt, 107, 123

Advanced Composition Explorer (ACE), 127, 155, 168, 198

Advanced Technology Solar Telescope (ATST), 202, 205

Albedo, 174, 179, 190

Antarctic, 33, 191–92; Antarctic Circle, 5, 42

Arctic, 33; Arctic Circle, 5–6, 42, 85; arctic ice, 180, 186, 188

Aristarchos of Samos, 2

Aristotle, 2, 61

Asteroseismology, 57

Astrology, 7–8

Atmosphere, loss of, 124

Aurora: colors, 159–60; electrojet, 161; emission, 84–86, 150; northern and southern lights, 85, 117–18

Babcock, Harold and Horace, 37, 42, 89

Babcock Model, 89–90

Bartels, Julius, 125, 159

Bartels diagram, 125

Beryllium-10, 86, 191

Birkeland, Kristian, 117–18

Bremsstrahlung radiation, 143

Bunsen, Robert, 15, 16, 104

Butterfly diagram, 40, 52, 74, 75, 87–88, 91

Calendars, 3, 5

Carbon dioxide, 173, 177–78, 192

Carbon-dioxide cycle, 179, 190

Carrington, Richard, 75, 135, 158

Chitzen Itza, 4

Chlorine-37, 86, 191

Climate, 81, 173, 175–78, 180–82, 205; and global warming, 184–92; models, 190–92

Comets, 117–19

Compass, 30, 32–34, 39

Conduction, thermal, 110, 112, 114–15, 134

Constellations, 7–8

Convection, 25, 41, 65–66; convection cells, 67

Convection zone. See Sun, structure of

Copernican Revolution, 2–3

Copernicus, Nicolaus, 2, 61

Coriolis effect, 41–42

Corona. See Sun, structure of

Coronagraph, 101–3, 147; occulting disk, 147

Coronal dimming, 133

Coronal heating problem, 101, 112–13, 115, 196; energy equation, 114

Coronal hole (*Koronale Löcher*), 90, 107–8, 112, 125; as origin of solar wind, 121–27

Coronal mass ejection (CME), 131–33, 145–50, 199; three-part structure, 146; prediction of, 156–58, 167–68, 197

Coronium, 113

223

Corpuscles, 117, 125
Cosmic rays, 86, 128–29, 192

Dalton Minimum, 83, 180
Davis, Raymond, 24, 54–55
Descartes, René, 14
De Vries, Hessel, 87
De Vries cycle, 86, 87
Differential rotation, 39–41, 51, 93
Dynamo: alpha effect, 41; omega effect, 41, 73; solar, 39–41, 93–94, 196; terrestrial, 33

Earth, 30–32, 173–74; changes in orbit, 175–76; life on, 1, 14, 173, 178; physical properties, 18. *See also* Climate
Eclipses, 101–2
E-corona, 106
Electrical blackouts, 14, 161, 163
Electromagnetic spectrum, 105, 133, 138, 182
Emission lines, 15, 23, 104, 136; in prominences, 148
Epicycles, 2
Equinoxes, 4–5
Eratosthenes, 6
Evershed, John, 71
Evershed Effect, 71
Extrasolar planets, 202–3; detection methods, 203–4

Fabricius, Johannes, 36, 60, 191
Faculae, 63, 70, 81, 184
Faint Sun paradox, 190; brightening Sun, 189, 191
Faraday, Michael, 34, 39
F-corona, 106
51 Pegasi, 203
Filament, 110–11, 133, 148–49. *See also* Prominences
Five-minute oscillations, 43–44, 49, 65
Fraunhofer, Joseph, 15, 104
Fraunhofer lines, 15–16, 106
Free radicals, 166

Galileo, Galilei, 8, 14, 60
Geographic north, 31–32
Geomagnetic indices, 92, 95, 158, 169
Geomagnetic storms, 81–82, 125, 135, 157–58; association with prominences, 148–49; recurring, 159
Geomagnetic tail, 32
Geostationary Operational Environmental Satellites (GOES), 140–41, 155
Gilbert, William, 30, 33, 34
Giovanelli, Ronald, 110, 137
Glacials, 176. *See also* Ice ages
Gleissberg Cycle, 83, 86, 87; Gleissberg, Wolfgang, 87
Global modes, 47, 50, 52
Global Oscillation Network Group (GONG), 47–49, 197
Global positioning system (GPS), 31, 162, 163
Global warming, 175, 178, 184–89, 192; average global temperature, 178, 186, 190; human activity (anthropogenic), 174, 187, 191
Golden Record, 129
Google Maps, 164
Gore, Al, 188
Gough, Douglas, 49, 190
Granulation, 25, 38, 62, 63, 65–66
Great Famine, 84
Great Storm of 1989, 160–63
Greenhouse effect, 177–79, 190
Greenhouse gases, 173, 177–78, 187–89, 192
Ground level event (enhancement), 135, 146

Hale, George Ellery, 36–37, 68, 148
Hale Magnetic Cycle, 88–89
Hale-Nicholson Polarity Law, 40, 75–76, 80, 88, 90
Hallstatt Cycle, 87
H-alpha radiation, 110–11, 136–38, 140
Harriot, Thomas, 35, 60, 191
Helicity, magnetic, 76
Heliopause, 99, 115, 197, 201

Helioseismology, 43–44, 48, 196–97; local, 52–53; nodal lines, 46–47; observations, 47–49, 51, 56

Heliosphere, 29, 115, 128–29, 201–2

Heliospheric current sheet, 109

Helium: discovery of, 104; in nuclear fusion, 21–22, 25–26, 54, 190; in the solar wind, 121, 124

Helmet streamer, 90–91, 108–9, 123

Hemispheric Helicity Rule, 76–77, 80–81

Herschel, Sir William, 19

Hinode, 198–99

HK Project, 96

Hodgson, Richard, 135, 158

Holocene era, 176

Homestake Goldmine, 55

Hydrostatic atmosphere, 113–14

Ice ages, 175–76, 179; glacials and interglacials, 176

Industrial revolution, 21, 191

International Panel on Global Climate Change (IPCC), 187–88

Interplanetary medium, 9, 141, 145; dust, 106; magnetic field (IMF), 108, 120–21

Interstellar medium, 99, 118–19, 128, 201

Ionosphere, 33, 156, 158, 161; disturbances in, 14, 135, 162–63

Janssen, Pierre Jules, 104, 148

Jet stream, 50, 52

Joy, Alfred, 76

Joy's Law, 40, 76, 89

K-corona, 106

Kepler, Johannes, 2–3, 117; laws of planetary motion, 3

Kepler Mission, 57, 204

Kiepenheuer, Karl Otto, 149

Kirchhoff, Gustav, 15–16

Kirchhoff's laws, 16

Kyoto Protocol, 188

Leighton, Robert, 43, 45

Limb darkening, 64–65

Little Ice Age, 81, 84, 180

Living with a Star, 56, 141, 201

Lockyer, Sir Norman, 104, 110

Loops, 100, 105, 109, 115; postflare, 100, 133, 142

Lunar base, 153, 155, 170

Lyot, Bernard, 101–2, 147

Magnetic field, general properties: lines of force, 34, 37, 121; magnetism, history of, 30–31

Magnetic field, of Earth, 31–33; magnetic declination, 32; magnetic poles, 32–33

Magnetic field, of the Sun, 23, 29, 34–37, 90; bipoles, 34, 37–38; buoyancy, 73, 89; magnetic carpet, 37–39, 196; closed field, 135; decay of, 38; dipoles, 89–91; magnetic flux tube, 70, 73; magnetic neutral line, 109, 111; open field, 120–121, 126, 156; poloidal field, 41, 93; reversal of, 41, 83, 88–89, 94, 196; of sunspots, 67–69, 72–73

Magnetite, 30

Magnetograph, 37

Magnetopause, 32

Magnetosphere, 30–32, 124–25, 161–62

Mars Exploration, 153–55; vision for space exploration, 170–71

Maunder, Edward Walter, 83

Maunder Minimum, 83–84, 91, 180–81

McIntosh, Patrick, 68–69, 149, 156

Medieval Maximum, 83, 92

Meridional flow, 52, 93, 94

Methane, 177, 178, 192

Milankovitch, Milutin, 176

Milankovitch cycles, 176

Mount Wilson Observatory, 36, 74, 75, 96

Mount Wilson Sunspot Classification, 68

M regions, 125, 159. *See also* Coronal hole

National Space Weather Program, 154

Neutrinos, 24, 53–55, 56

Newton, Harold, 135, 149

Newton, Isaac, 3, 15, 17; *Principia*, 14, 17

Nicholson, Seth Barnes, 75, 80

Nuclear fusion, 21–24, 53, 55; p-p chain, 54

Oort Minimum, 83

Ørsted, Hans, 33, 39

Ozone, 173, 178, 181, 184

Parker, Eugene, 118, 133

Parker's Model, 118–20

Parker Spiral, 120–21, 132, 148, 157

Photosphere. *See* Sun, structure of

Pinhole projector, making a, 60

Planetary orbits, 3, 17, 203, 205; orbital period, 17

Planetary Systems, 1, 203–5

Planetary transit, 17, 203–4

Plato, 2

Pleistocene Era, 176

Prominences, 100, 105, 110–12; active region, 111; *disparition brusques*, 148; eruptive, 132, 137, 148–49; quiescent, 111, 138. *See also* Filament

Ptolemy, 2

Quaternary Period, 176

Quiet solar network, 38, 63, 107

Quiet Sun, 42, 63, 67, 107, 112

Radiation, 153, 161, 165–66; exposure (dose), 155, 166, 171; hazards, 154, 156, 162–63, 197; cancer, 165; radiation sickness, 166

Radiation belt, 128

Radiocarbon (Carbon-14), 83, 85, 191; radiocarbon dating, 86

Roentgen Equivalent Man (REM), 166

Scale height, density, 114

Scheiner, Christopher, 36, 60, 191

Schwabe, Heinrich, 73, 79, 82, 92

Scientific notation, 16

Shocks: in solar wind, 133, 145, 147–48, 168; termination, 129, 201

Sigmoids, 77, 109–10

Single event upset (SEU), 154, 162

Snowball Earth, 179–80

Solar activity, prediction and forecasting, 154–56, 167–69, 197

Solar and Heliospheric Observatory (SOHO), 48, 132, 141

Solar constant, 19–20, 174–75, 182, 189–90

Solar cycle, 40, 73–74, 79–80, 82–86; initiation of, 89–91; prediction of, 91–95; terrestrial impact, 81

Solar day, 7

Solar Dynamics Observatory (SDO), 56, 201–2

Solar energetic particles (SEP), 154

Solar flares: definition of, 134; CME association, 131–33; energy release, 137–39, 144; nonthermal radiation, 141–44; optical classification, 140; prediction of, 150, 155–56, 167–68, 197; time evolution, 134–36; white-light flare, 135; X-ray classification, 140–41

Solar irradiance, 181, 202; total solar irradiance (TSI), 84, 183

Solar luminosity, 20–21, 174–75, 190

Solar maximum, 81, 90, 122, 123

Solar minimum, 90, 93, 108–9

Solar nebula, 21

Solar neutrino problem, 24, 44, 53–55

Solar Orbiter, 56, 202

Solar oscillations, 44–45; "breathing mode," 47

Solar proton events (SPE), 154, 157, 165

Solar radiance, 181

Solar radiation, 138–39, 155; Calcium II, 96; EUV, 60, 102, 105, 133; gamma-rays, 133, 142, 144; hard

X-rays, 141–44; radio emission, 92, 135, 154; soft X-rays, 84–86, 108, 140–41; UV, 63, 184

Solar rotation, 39, 50–52; frequency splitting, 51

Solar spectrum, 15, 106; spectrograph, 36–37

Solar storms, 10, 85, 150, 153, 162

Solar System, 2–3, 17, 21, 99, 128–129, 170, 197

Solar Terrestrial Relations Observatory (STEREO), 102, 141, 199, 201

Solar wind, 13–14, 119; affect on planets, 124; and the magnetosphere, 32–33, 157–58; fast and slow streams, 121, 123, 159; fast and slow wind, 108, 121–23, 127; mass loss, 112, 124; source of, 125–27

Solstice, 2–5

Sound waves, 43–46

Space Age, 31, 132

Spacecraft drag, 92, 156, 162

Space stations: Skylab, 145, 149; International Space Station (ISS), 158, 165, 170

Space weather, 131, 147, 150, 154, 163; alerts, watches, and warnings, 167, 169

Spicules, 100, 107, 126

Sporer, Gustav, 40, 83

Sporer Minimum, 83–84, 92, 180

Sporer's Law, 40, 75, 88

Standard model, particle physics, 55

Standard solar model, 48

Stars: activity (stellar), 10, 95–96; chromospheres of, 96; cycles, 95–97; evolution, 22, 24; main sequence, 25, 203; oscillations, 57; spectral type, 16, 95; spots, 77; Sun-like, 18, 22, 96–97, 196

Stefan-Boltzmann law, 174

Stellar fingerprinting, 15–16

Stonehenge, 4

Streamer belt, 90, 109

Sun, mythology, 3–4

Sun, navigation, 6–7, 30; cross-staff, 6; sextant, 6

Sun, structure of: atmosphere, 23, 99–102, 106, 112, 115; chromosphere, 100, 105, 110, 112, 196, 201; convection zone, 25, 39, 196; core, 20–22, 25, 53–55, 190; corona, 100, 102–3, 106, 112–13, 196; disk, 56, 62, 64–65, 71; limb, 62, 64; photosphere (surface), 61–63, 65, 77; radiative zone, 23–25, 39, 52; transition region, 100, 105, 110

Sun, timekeeper, 5–7; Harrison, John, 7; shadow clocks, 6

Sunspots: bipolar, 68–69, 73; cluster model, 70; delta-spots, 69, 137, 160; following, 40, 69, 73, 75–76, 89; leading and preceding, 40, 69, 73, 75–76, 80, 88–89; penumbra, 67–68, 70–72; pores, 67, 70; rotation, 72; sunspot area number, 74–75; sunspot groups, 69–70, 75, 91, 160; umbra, 67–68, 71; unipolar, 68–69

Supergranulation, 66, 149

Tachocline, 39, 50, 52

T-corona, 106

Temperature minimum region, 62–63

10.7 cm microwave emission, 84, 92, 95

Termination shock, 129, 201

Thales of Miletus, 30

Thomson, J. J., 103

Thomson scattering, 103–4, 106

Toroidal field, 41, 91, 93

Total solar irradiance (TSI), 183–84

Transition Region and Coronal Explorer (TRACE), 141, 198

Tropical year, 5, 7

Ulrich, Roger, 43, 45–46

Ulysses, 121–23, 127, 141, 197

Voyager spacecraft, 115, 129, 197

Waldmeier, Max, 68, 107
Waves, properties of, 44–45, 134
Wien's Law, 103
Wilson, Alexander, 67
Wilson depression, 67
Wolf, Rudolf, 74
Wolf Minimum, 83, 180
Wolf number, 74

Zeeman, Pieter, 36
Zeeman splitting, 36, 50
Zodiac, 7–8
Zodiacal light, 106
Zurich number (Waldmeier number), 68, 79

About the Author

DAVID ALEXANDER is currently the Andrew Hays Buchanan Associate Professor of Astrophysics at Rice University in Houston, Texas. He received a Presidential Early Career Award for Scientists and Engineers in 2004 and was appointed a Kavli Frontiers Fellow by the National Academy of Sciences in 2006. He is a co-investigator on the NASA STEREO mission and an Associated Scientist on the NASA Solar Dynamics Observatory mission, to be launched at the end of 2008. Professor Alexander serves as an associate editor for the *Journal of Geophysical Research–Space Science* and has also served on several national advisory committees for NASA, the American Geophysical Union, and the Solar Physics Division of the American Astronomical Society.